Recording Orchestra and Other Classical Music Ensembles

Recording Orchestra and Other Classical Music Ensembles explores techniques and methodologies specific to recording classical music. Whether the reader is a newcomer or a seasoned engineer looking to refine their skills, this book speaks to all levels of expertise and covers every aspect of recording symphonic and concerto repertoire, opera, chamber music and solo piano.

With a focus on the orchestra as an instrument and sound source, *Recording Orchestra and Other Classical Music Ensembles* features sections on how to listen, understanding microphones, concert halls, orchestra seating arrangements, how to set up the monitoring environment and how to approach recording each section of the orchestra. Offering concise information on preparing for a recording session, the role of the producer and mixing techniques, whilst a "quick-start" reference guide with suggested setups also helps to introduce the reader to the recording process. Online Instructor and Student Resources, featuring audio and video examples of various techniques, further reinforces the concepts discussed throughout the book.

This new edition has updated and expanded material, including new chapters on classical crossover projects, film score recording and immersive/3D recording and mixing, as well as a number of new case studies, making this an essential guide for students, researchers and professionals recording classical music.

Richard King is a multi-Grammy Award winner who specializes in recording classical, jazz and film score music. He is Associate Professor at the Schulich School of Music of McGill University in Montréal, where he teaches in the Sound Recording area. A long-standing member and Fellow of the Audio Engineering Society, Richard is a regular convention presenter and panelist, and he is frequently invited to speak at various university-level recording programs around the world.

"King's *Recording Orchestra* is nothing short of indispensable for anyone interested in the art of ensemble recording (…) it's probably the finest book on recording technique I've read in a decade."

Mike Senior, *Sound On Sound* Magazine on the 1st Edition

AUDIO ENGINEERING SOCIETY PRESENTS …

www.aes.org

The MIDI Manual, 4th Edition
A Practical Guide to MIDI within Modern Music Production
David Miles Huber

Digital Audio Forensics Fundamentals
From Capture to Courtroom
James Zjalic

Drum Sound and Drum Tuning
Bridging Science and Creativity
Rob Toulson

Sound and Recording, 8th Edition
Applications and Theory
Francis Rumsey with Tim McCormick

Performing Electronic Music Live
Kirsten Hermes

Working with the Web Audio API
Joshua Reiss

Modern Recording Techniques, 10th Edition
A Practical Guide to Modern Music Production
David Miles Huber and Emiliano Caballero

Recording Orchestra and Other Classical Music Ensembles, 2nd Edition
Richard King

For more information about this series, please visit: www.routledge.com/Audio-Engineering-Society-Presents/book-series/AES

Recording Orchestra and Other Classical Music Ensembles

Second Edition

Richard King

NEW YORK AND LONDON

Designed cover image: Église Saint-Jean-Baptiste/Saint-Jean-Baptiste Church, in Montreal.
Photo Credit: Owen Egan.

Second Edition published 2025
by Routledge
605 Third Avenue, New York, NY 10158

and by Routledge
4 Park Square, Milton Park, Abingdon, Oxon, OX14 4RN

Routledge is an imprint of the Taylor & Francis Group, an informa business

First edition published by Routledge 2017

Library of Congress Cataloging-in-Publication Data
Names: King, Richard (Richard Lemprière) author.
Title: Recording orchestra and other classical music ensembles /
Richard King.
Description: Second edition. | New York: Routledge, 2024. |
Series: Audio Engineering Society presents |
Includes bibliographical references and index.
Identifiers: LCCN 2023041574 | ISBN 9781032349435 (hardback) |
ISBN 9781003319429 (paperback) | ISBN 9781003324607 (ebook)
Subjects: LCSH: Sound recordings–Production and direction. |
Sound–Recording and reproducing.
Classification: LCC ML3790 .K4648 2024 | DDC 784.2/149–dc23/eng/20230905
LC record available at https://lccn.loc.gov/2023041574

ISBN: 978-1-032-34943-5 (hbk)
ISBN: 978-1-003-32460-7 (pbk)
ISBN: 978-1-003-31942-9 (ebk)

DOI: 10.4324/9781003319429

Typeset in Times New Roman
by Deanta Global Publishing Services, Chennai, India

Access the Instructor and Student Resources: https://routledgetextbooks.com/textbooks/9781138854543

In memory of my dear father, Dr. Hubert W. King.

Contents

Acknowledgments

I would like to thank my wife Isabelle and my son Léo for their love and support. To my parents Ann and Hu' – thanks for everything, but especially for supporting my various musical and audio endeavors throughout my youth and early adult years. Thanks to Hannah Rowe at Taylor and Francis for her interest in the second edition and for soliciting and managing feedback from users of the original book. For their help in the editing process, thank you to Emily Tagg and Frances Tye, and to Francis Rumsey for his role as technical editor for the first edition. Regarding the audio examples, I would like to thank composer John Adams and Nonesuch Records; Jim Gray and the Brass Band of Battle Creek; Maestro Alexis Hauser and the members of the McGill Symphony Orchestra (also featured on the cover); and soloists Meagan Milatz (piano) and Victor Fournelle-Blain (violin), pianist Kimihiro Yasaka, and violinist Kate Maloney and her string quartet. Thanks to recording engineers Paul Hennerich, Jack Kelly, Ben Ewing, Marcelo Saurez and Jordan Strum, and finally Denis Martin for his contribution as "drone pilot" for some of the video segments. Thanks to Fei Yu for all of her work and expertise in reviewing the Chinese translation of the first edition.

Lastly, a **very special thank you** to those directly responsible for my education in music and audio. This list is meant to be in chronological order but there is some overlap: Priscilla Evans, Alan Gaskin, Jim Faraday, Dennis Farrell, Geoff Doane, Peter Cook, Wieslaw Woszczyk, Tim Martyn, David Smith, Buddy Graham, Steven Epstein, Shawn Murphy and George Massenburg.

About the Author

Credit: Peter Matulina.

Richard King was born in the UK and grew up on the east coast of Canada. He completed his Bachelor of Music degree in percussion at Dalhousie University in Nova Scotia, and his Masters of Music in Sound Recording at McGill University in Montreal. After two seasons at the Tanglewood Music Center (summer home of the Boston Symphony), he was hired by Sony Classical Productions in New York, and in less than a year he was assigned a project as engineer with the Los Angeles Philharmonic Orchestra and conductor Esa-Pekka Salonen. Over the next 15 years, based at Sony Music Studios in New York City, he had the great fortune of working with many famous orchestras and conductors worldwide, and with top soloists including Yo-Yo Ma, Joshua Bell, Chick Corea, Béla Fleck, Hilary Hahn and Renée Fleming, and composers John Adams, Wynton Marsalis, Edgar Meyer and Billy Joel. Richard has also been involved with many classical crossover, Broadway and jazz projects as well as a long list of film scores, and has recorded and mixed music for Eric Clapton, the Punch Brothers, Jerry Douglas, Ben Folds and James Taylor.

Over the years, Richard has garnered multiple Grammy awards over many categories, including Best Engineered Album, Classical *and* Non-Classical, Best Orchestral Performance, Folk Album, Score Soundtrack, Musical Theater, Classical Crossover, and a Latin Grammy for Best Instrumental Album. In 2009 Richard accepted a position as Associate Professor in the Graduate Program of Sound Recording at the Schulich School of Music of McGill University in Montreal. He is also a regular guest lecturer at various recording programs around the world. A Fellow of the Audio Engineering Society and long-standing member, Richard is a regular convention presenter and publisher, student mentor and workshop panelist, and a member of the Technical Committees Spatial Audio and Recording Technology and Practices. His research interests include music mixing methodologies and immersive/3D audio production. Richard is also a member of the Acoustical Society of America and the Producers' and Engineers' wing of the Recording Academy (NARAS).

Preface to the Second Edition

This second edition of *Recording Orchestra* comes into print around seven years after the first run. During that time there have been many changes in the industry and to the technology used; however, many elements remain the same, such as the microphone types used and the techniques behind their careful placement during the recording process. Additional material in this printing includes several new chapters on topics that I had decided not to include in the first edition: specifically, those on classical crossover projects and film score recording, immersive/3D recording and mixing, along with two chapters of case studies drawn from actual recording sessions. Several other full chapters and sections have been added where I honestly overlooked their inclusion first time around. These include chapters on woodwind and brass quintet as well as on chorus with piano, and a more complete discussion of wind symphony and brass band.

One major structural change is that all of the "Try This on Stage" text boxes have been reorganized as more formal exercises, and a full List of Exercises can be found just before page 1. Many of the original chapters now have additional figures showing certain techniques and microphone placements more clearly.

From the First Edition Preface

The goal of this book is to get directly into the subjects of listening, microphone placement and the aesthetics of perspective and balance in classical music recording and mixing. Throughout the pages of this book, various methodologies for classical music recording are discussed, and certain specific techniques are demonstrated in detail, with clear explanations as to why they might be adopted as successful solutions. Starting positions for microphone placements are suggested so that the reader has an initial beginning location on stage that they may adapt and adjust to suit their personal preference. Suggestions for how to refine the resulting audio capture are laid out in an easy-to-follow manner, so that a successful recording can be made in practically any situation.

Many chapters feature a section or two with specific exercises for evaluating sound and experimenting with microphone placements and mixing techniques. The audio examples on the publisher's companion website, (https://routledgetextbooks.com/textbooks/9781138854543), in conjunction with the Quick Start Guides in the Appendix, help make this book as "hands-on" as possible. I encourage all readers to experiment for themselves, and hope that the techniques laid out in the following pages aid them in developing their personal methodologies.

Richard King

Exercises

Part I

Getting Started

Chapter 1

Introduction

Throughout the industry, there exists a broad range of accepted practices and styles of audio production and engineering. This highly subjective field is driven by the individual preferences of practitioners, along with consumer expectation. In music production one engineer might prefer a "bright" mix, while another might choose a "warmer" or darker presentation, and both approaches may be entirely valid. I like to draw on metaphors from the culinary discipline, as both audio recording and preparing food are executed "to taste". For instance, what exact measure constitutes a "bright" mix? This is a subjective value, just like the questions: How salty or spicy is the stew? How much garlic is too much garlic, or how much added reverb is too much reverb? My point here is not that garlic and reverb have much to do with each other, rather that these questions of preference in audio, as with food, will yield a wide range of responses when presented to a large group of people.

Newcomers to the field of audio engineering will begin to develop a certain personal aesthetic over time, which resides within a "window of acceptability" that should satisfy most listeners. A general style or methodology must be developed that is not "bland", but at the same time, not too salty or spicy. This may take years to hone and refine, but it is a natural progression. These very basic sonic attributes are mentioned here only as examples, while more specific points will be addressed throughout the book. Except for a few overlapping procedures, most recording techniques used in classical music production are quite different from those used in popular music. As such, the text will follow a scope that is more relevant to the classical music genre. That isn't to say that these techniques cannot be adapted to pop music recording. I have had a certain measure of success applying these same techniques to jazz and pop projects, albeit with a few modifications along the way. The subsequent chapters are meant to give newcomers a head start and some basic guidance as they settle in on preferred and individualized methods of audio recording.

1.1 Schools of Thought on Orchestral Recording

Several different approaches exist on how to record classical music. These various techniques have been in use for decades, and each is valid in its own way. An overview of a few examples follows in the next pages. The takeaway is that any technique can work well as long as the engineer truly understands the functionality of that particular approach, and can properly master it, thereby producing a natural result. I forget who said first (it wasn't me, although I say it often) that in classical music recording, we are "trying to create the *illusion* of reality"; to that end, we may break the rules along the way, but as long as we fully understand those rules, we can compensate for any unnatural effects of each "transgression".

DOI: 10.4324/9781003319429-2

1.2 Creating the Overall Picture from One Perspective

Through this approach, the goal is to carefully position a main microphone system so that a satisfactory general balance of the ensemble is captured along with an appropriate amount of room sound from the hall, otherwise known as the direct-to-reverberant ratio [1]. This set of main microphones is usually complemented with other supporting or "spot" microphones, which are placed closer to the sources and introduced into the mix at a lower level so that the main sound and balance is primarily that of the principal pickup. A great advantage to this approach is that the resulting balance should be quite close to that which the conductor hears when standing on the podium. This approach offers a natural perspective, whether it is at a more intimate or a more distant placement, with purity in signal, as long as additional elements are introduced at a conservative level. The disadvantage of this approach is that more time is required, and more experience is necessary to optimize the position of the microphones. Careful listening to a dress rehearsal or soundcheck "over the microphones" is required to ensure that the best placement has been realized.

1.3 Combining Two or More Main Microphone Systems

Utilizing a pair or "system" of main microphones together with a more distant, and normally more widely spaced pair of microphones is another common approach. In this case, the two (or more) pairs can be combined at varying levels to achieve an appropriate sum of clarity and reverberation or "bloom" from the hall. For recordings in which the engineer is unsure of the characteristics of the hall, or when very little time is afforded for a soundcheck, this may be an informed approach. One risk with this method is that the result may simply yield the collective perspectives of "too close" and "too far", rather than a well-blended, natural-sounding presentation. Also, it should be kept in mind that the overall sound might become less defined, as more and more main microphone systems are integrated at roughly equal levels. This may, however, be a positive result in terms of blend.

1.4 Using Close Microphones as the Principal Audio Capture

This is an older technique that was commonly used for many years and is still used by a few engineers and recording teams. The idea is to capture each section of the orchestra as separate elements, combining the signals at appropriate levels to achieve a suitable balance, and then a pair (or more) of room microphones is introduced to fill in the sound of the room, thereby "gluing" together the entire presentation. This technique allows for more control and flexibility in the resulting sound, and therefore might be considered an appropriate technique for live recording where no time has been afforded to check the sound and balance before the concert. The downside to this approach is that more work will be required to create a "natural" sound, in terms of perspective and blend.

1.5 Other Classical Music Ensembles

As with orchestra recording, string quartets, piano trios and other chamber groups can be discussed in the same manner. The recording technique might be based on an initial assessment of a main pair or system, with a great deal of time spent moving the microphones up, down, in and out, before considering the addition of any supporting microphones. Alternatively, a more controlled approach may be preferred – balancing a series of microphones placed close to the instruments and supplementing with a pair of room microphones. Recording solo instruments such as piano or violin can be a very complex or incredibly simple process, depending on the technique implemented and the acoustical properties of the recording venue.

It is no wonder that aspiring audio engineers can become quite confused and even discouraged as they experiment on their own. It is the intent of this book to provide a starting point for recording each type of ensemble, and to offer clear and simple guidance on how to make well-balanced, commercial-quality recordings with repeatable results.

Various techniques and approaches will be discussed and evaluated in an objective manner, so that the reader may experiment with all the available tools and have the chance to form their own conclusions regarding the relative success of each approach. All of the suggested techniques come from real-life situations and are designed either to leave certain options open in post-production, or simply to help carry an engineer through a difficult session. The methods are all practical and well proven for surviving in the field and making high-quality recordings on a consistent basis.

1.6 Chapter Summary

Although there are many ways to approach classical music recording, it is the end result that is truly important – a mostly accurate capture of a large ensemble that generates a compelling listening experience. I say "mostly" because in certain cases we may be trying to present an enhanced experience, with a "larger-than-life" presentation. In other words – enhanced bass, a wider image, extra reverb, clarity in low-level details, and slightly exaggerated solo levels and soloist balances are all important areas of attention in modern recording. This approach will help solve the problem of how to capture the experience of hearing a great orchestra in a wonderful hall, so that it translates as best as it can over good-quality loudspeakers or a decent pair of headphones. Anyone who has lived the exhilarating, visceral experience of standing a few feet behind a conductor on the stage of a large orchestra playing at full volume will understand how difficult it is to try and electronically recreate the event, even with a good selection of professional microphones and a decent playback system. For me, the experience came during a recording session in October of 1992, with Esa-Pekka Salonen conducting the Los Angeles Philharmonic Orchestra in Royce Hall, UCLA. I will never forget this incredibly vivid and breathtaking moment.

Reference

1. Beranek, L.L. (1986). *Acoustics*. New York: Acoustical Society of America.

Chapter 2

How to Listen

Loudspeakers, Headphones and Listening Rooms

While it may seem obvious, it should be noted that the first step in learning how to make a recording is to learn how to listen. The popular audio moniker "Golden Ears" is actually a misnomer, as it implies that an audio specialist is born with or has somehow developed superhuman hearing. Almost everyone starts out with the same level of hearing ability or "sensitivity" – the key is to *train the brain* so that it can most effectively decipher the signals received by the ears. This chapter sets the groundwork for preparation as an audio practitioner and begins with a discussion of sonic characteristics of sound, followed by exercises in listening to existing recordings, assessing the listening room, and evaluating audio over live microphones at a recording session or the dress rehearsal of a concert event.

2.1 Learning to Listen

When I was young, I remember my fifth-grade teacher asking our class to close their eyes and take note of the sounds around us, as an exercise in aural awareness. I also remember thinking that it was a waste of time, but fortunately since the age of eight I have refined my opinions on many subjects, including music and sound.

Most humans prioritize vision as the dominant sense, while the incredibly powerful sense of hearing tends to be underutilized [1]. Think about it – the normal field of peripheral vision extends to a total of 180° in the lateral plane and less in the vertical dimension, and each eye comes equipped with its own blind spot. The ears, however, capture sound from all around us, with increased sensitivity in high frequencies above ear level in front of and to the sides of the listener. Of course, there is difficulty in localizing sound directly behind the head, but as children we learn how to do this through practice and are therefore able to adapt. As a warning system, the ears can help us avoid stepping into unseen traffic and can alert us when a pizza should be removed from the oven, as we hear the timer ringing while relaxing in the next room. Many subtle aural cues go completely unnoticed on a daily basis by normal listeners. The sound is "heard" but not "observed". One of the greatest benefits of hearing is the profound and fascinating manner in which we can be emotionally affected by listening to music, whether it is being performed live or electronically reproduced.

It should be noted that the function of listening and the efficiency of the hearing system are affected by many external factors. Alcohol and caffeine, lack of sleep and loud sounds all affect the sensitivity of the ear in varying degrees. The listener's objectivity and general ability to evaluate sound (judgment) are also important influences on the auditory system.

DOI: 10.4324/9781003319429-3

2.2 Characteristics of Sound

The first step in learning how to listen to recorded music is learning to identify the various sonic characteristics of the program. Each of these traits must be taken into consideration separately as an overall impression of the recorded sound is realized. Measures of timbre include frequency, amplitude and to some extent distortion, while spatial metrics describe image and location, source width, height and depth. Dynamic range and dynamic contrast are less easily understood, and commonly require the most training. For more precise discussions of the attributes of sound, I would recommend some further reading [2, 3]. The European Broadcasting Union (EBU) suggest a similar, standardized group of defined terms in their document EBU Tech 3286 [4]. The most basic criteria that might be analyzed are as follows:

Frequency: The sonic range from low to high as it relates to pitch, rather than volume or level. "Bass" and "treble" sounds correlate to low- and high-frequency signals and are measured in hertz (Hz).

Amplitude: The level or intensity of a signal. Volume is commonly used as an expression of amplitude in sound – measured in decibels (dB).

Image/Perspective: In stereo audio reproduction, image is the presentation of signal across the lateral soundstage, from the left loudspeaker to the right. Source location and source width focus on where the source "appears" between the two loudspeakers. Perspective of image should also be considered here in terms of width, depth and overall spaciousness. How close or distant is the overall presentation?

Stage placement (location): For orchestra, this is the physical position of each section on the stage. Are the violin sections split to the left and right sides? Are the French horns behind the woodwinds, or on the left side, opposite the rest of the brass? Where are the percussion, harp and celeste positioned? This should all be taken note of in the first moments of listening (or at each instrument entrance, according to the composition – you might have to wait for harp and percussion if they only play later in the piece).

Balance: The volume or level of each instrument or section in the orchestra as compared to the overall presentation. Balance can be assessed in many ways – string section balances, strings vs. brass, front half of the orchestra vs. back half, and so on. Section blend should also be evaluated as part of the balance. For instance, are the "first desks" of strings sitting near the conductor more prominent than other string players? For concerto recordings or pieces with voice, the balance of solo element vs. orchestral accompaniment must be observed. The overall result will be a combination of the conductor's efforts and the microphone placement and mix. The balance of direct sound vs. reverberation may also be considered here, although this is parameter that mostly affects perspective, so it is included under *Image*, above.

Dynamic range: The difference between louder and softer passages of musical program. This should be evaluated within movements, comparing various musical passages of different intensities, as well as louder and softer movements of a complete symphony.

Dynamic contrast: The difference between sustained musical program and the transient peaks associated with the onset or attack of a note. This parameter is more relevant in pop music, where compression or dynamic range control as a creative production tool is more widely used. In most classical music recordings, the dynamic contrast is generally kept the same as captured at the microphones; however, reduced dynamic "range" can be the result of a distant microphone placement rather than overuse of compression – so it is important to make note of this attribute when evaluating any recording.

Naturalness: The measure of how closely the recording resembles the sound of an actual orchestra performing in an acoustic space, for example a hall or church. Various aspects will affect this particular parameter. A low rating in this category might be the result of the perspective being too close or distant, or might occur if the image is too narrow or the balances are dramatically off. Also, if too much signal processing has been used, the result will be a less authentic presentation. Lack of ambience or reverberation can take away from a natural-sounding result, whereas too much artificial reverberation or the use of an inappropriate reverb program can also diminish the naturalness of the presentation.

Distortion and noises: Overloading the input of a microphone preamp and clipping of a digital signal during loud musical passages are two of the most common and noticeable examples of distortion in classical music recording. Quite often, however, there are distorted signals present in the softer passages of a piece as well, possibly due to a faulty converter or some other subtle problem. One must listen for distortion and noise in both loud and soft passages, noting acoustic and electronic artifacts. Ticks, clicks, thumps, chair noises and rumble from traffic and air conditioning, as well as hum and buzz from power transformers and lighting systems and dimmers, must all be evaluated as to their severity, as they detract from the quality of the listening experience. Noises might be divided into two groups – acoustic artifacts captured by the microphones, and electronic issues such as a click or pop caused by a faulty word-clock configuration.

2.3 Listening to Existing Recordings

The novice audio engineer needs to "train their brain" by listening to a multitude of existing commercial recordings so that they become familiar with an audio image of what might be considered an "average" presentation of an acoustic event. The engineer needs to have an opinion going into a recording project and will benefit from having a reference recording on hand, so that they have somewhere to start. Using existing recordings is the best practice in learning how to listen.

With experience, it should be possible for the listener to quickly take a "snapshot" of all the audio aspects of a recording. These aspects include perspective, image width and depth, timbral balance, sectional balance within the orchestra, presence and perspective of each section, dynamic range and contrast, and noise issues such as distortion, rumble (traffic noise) and hum. It is recommended that the listener quickly form an overall opinion before the auditory system is affected by adaptation and the timbral and spatial dimensions start to sound "balanced". Longer periods of listening can be beneficial for certain aspects of audio which change over time, such as dynamic contrast, but the steady-state attributes of timbre or image width should be assessed quickly, and first impressions will be the most accurate. Taking notes while listening is a helpful tool, as it allows the listener to compare early impressions with those developed over longer listening periods. Additionally, the effect of listener adaptation will be more apparent as written observations are seen to change over time.

In the analysis chart below (Table 2.1), it is suggested that a combination of text and images is best while making notes during the listening process. For frequency response this can be done by drawing a (solid) line from left to right, with changes in amplitude shown as higher or lower regions of the line. For example, a recording with lots of low-frequency buildup and very bright high strings might be drawn to look like this (Figure 2.1):

Table 2.1 Analysis Chart for Critical Listening

Descriptor	Meaning	Possible Answers
Frequency Response: Frequency vs. Amplitude	Overall, is it "even" or "flat"? Bright, Dark, Muddy, Thin? Good HF and LF extension? Possibly draw a graph …	
Image/Perspective	Describe the overall image. Wide? Narrow? Even from left to right, or is there a hole in the middle? Is the depth exaggerated?	
Stage Placement	Quickly identify the stage layout – how are the string sections sitting? Where is brass, harp, perc, etc.? Possibly draw a sketch …	
Balance	Describe the overall orchestral balance. Strings vs. brass, or front vs. back? Solo vs. orchestra?	
Dynamic Range/ Dynamic Contrast	How great is the difference between loud and soft sections of the piece? How great is the difference between attacks and sustained notes in the music?	
Naturalness	Does the recording represent a realistic event? Does the quality and amount of reverberation sound natural?	
Distortion and Noises	Are there any distracting artifacts such as clipping, rumble, hum or buzz from lights etc.?	

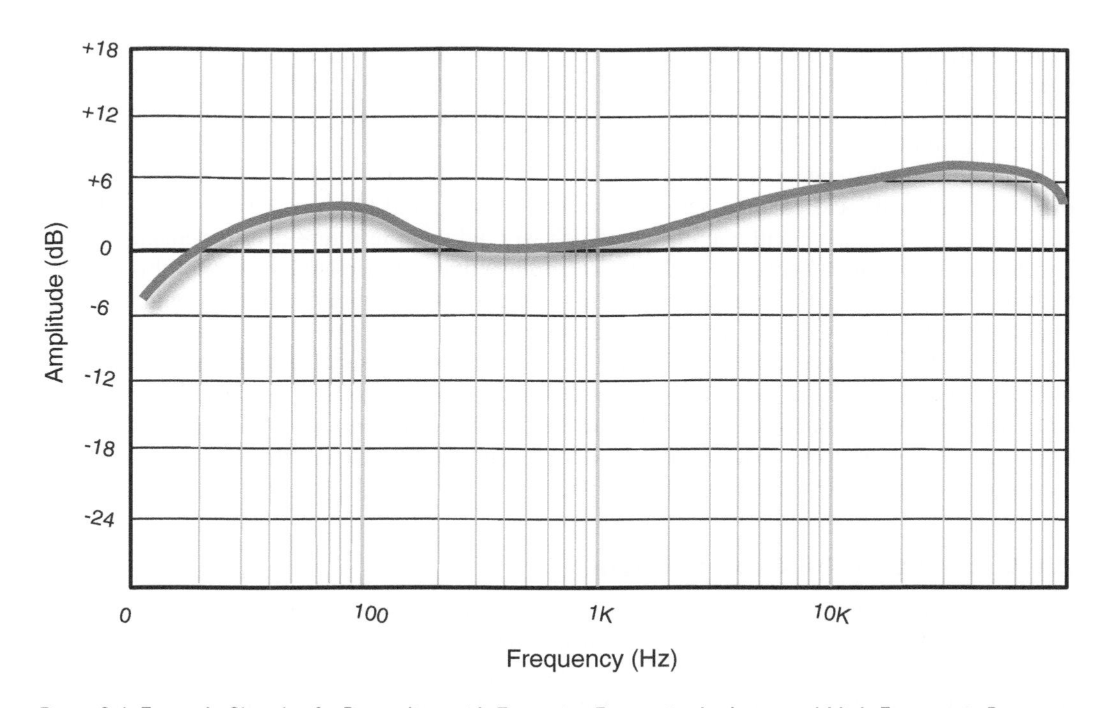

Figure 2.1 Example Sketch of a Recording with Excessive Energy in the Low- and High-Frequency Range

EXERCISE 2.1 LEARNING TO LISTEN

1. Choose any stereo orchestral recording, and while referring to Table 2.1, identify and evaluate the various sonic qualities.
2. Those who are more familiar with listening to popular music recordings might prefer to begin this process by listening to pop material, as the learning experience is equally valid. I would however encourage those same readers to jump directly into the deep end using orchestral recordings, since the focus of this book is on classical music recording techniques.
3. Subsequent chapters will be more relevant to those who have developed some confidence and specific expertise as evaluators of orchestral recordings. This table is also available online for download as a template at https://routledgetextbooks.com/textbooks/9781138854543/.

This analysis process can and should be extended to the study and comparison of multiple recordings of the same repertoire – for example, three different performances of the same Beethoven symphony recorded in different venues, with different orchestras and conductors. There will of course be some outlying variations that will distract the listener, such as tempos and interpretation, and quality of performance. As much as these musical attributes can affect the overall impression of the audio, the listener has to learn to ignore the musical differences while focusing on the analysis of the sonic characteristics of each recording. At this point it is best to keep the music score closed, so that the focus is on the sound and not the music. If something in the sound is hard to decipher, across all versions of the same piece, then the score can be a great help. Listening and analysis should be done using headphones as well, so that the brain can be trained to understand the differences between headphone and loudspeaker monitoring. Headphone listening is addressed in more detail in Section 2.6.

Over time, the listener will develop a sense of what constitutes a good recording in terms of overall presentation and balance. During this process, a natural bias toward a certain style will be developed. This will instill confidence, which will aid them in future decision-making and in developing a consistent recording aesthetic. Recording engineers need to be aware of this bias and treat it as such – a subjective preference. This is important to mention, since the engineer must remain open to change, and be able to adjust their own preference to suit the "taste" or natural bias of both the producer and the artist. A good example is the use of artificial reverberation. If they tend to use a great deal of "reverb" while mixing, they must be ready to adjust this preference when working with an artist or client who prefers a less reverberant overall sound. As discussed in Chapter 1, not everyone enjoys spicy food, and you won't make any friends by insisting that everyone at the table covers their food in hot sauce!

2.4 Listening to the Control Room

A proper assessment of the listening room should be made before any evaluation of "live audio" is considered – that is, the signal coming from the microphones in the hall or the studio. The engineer needs to know what is happening sonically in their monitoring environment in order to properly judge the sound and balance during a soundcheck or dress rehearsal. For instance, if there is a large bass buildup in the control room, yet the incoming audio from the stage sounds more-or-less even, one can infer that the resulting recording will be "bass-light", and adjustments should be made to correct for this deficiency (see Chapter 6). Of course, if a low-frequency buildup is also present when evaluating the live audio in the same control room, the recording will be more-or-less even, except that the live listening experience will be somewhat unsatisfactory unless something can be done to correct the buildup of sound in the listening room (see Section 2.7).

Even a decent control room with a more-or-less "flat" response (equal energy from low to high frequencies) will need to be carefully checked out before decisions are made on microphone placement. Microphone positions will always be subject to the influence of the listening conditions in the control room. There are many variables in play – loudspeaker choice, placement of the loudspeakers in the room, control room volume setting and the general acoustic characteristic of the room. Imagine a control room that is quite comfortable to work in but is slightly deficient in the midrange – even a small amount such as two or three dB will create a noticeable difference in a vocal balance with orchestra. This "minor" flaw could potentially lead the engineer to over-mix the soloist in a concerto recording. What if the sound in the control room is unusually live or dead? A very live control room may lead the engineer to place the main microphones too close to the source, and the resulting recording might sound too tight and dry in other listening rooms.

EXERCISE 2.2 EVALUATING THE CONTROL ROOM

1. Following the practice recommended in Section 2.3, play a few of your most familiar, well-studied recordings in the room where you plan to monitor during the recording and take some notes on what you hear.
2. These recordings should come from the same inventory that was used to train the brain. Mental notes are fine, but for novice listeners documenting written impressions is highly recommended, so that they can be referred to later on.

This is the best way to become comfortable with listening in the control room. I have three recordings that I audition in every listening room I use – even in rooms I have used quite frequently over the years. As you can imagine, I am now sick of hearing those recordings (and as such they will remain nameless), but I know them so well I can very quickly assess the space in which I will be critically listening to less familiar program thereafter. The idea is to eliminate variables in the process – evaluate an unknown space with known recordings, and then assess the sound of a new recording in a room that has now become a familiar listening space. Floyd Toole speaks of the "Circle of Confusion" in his book *Sound Reproduction*, wherein recording and mixing engineers complicate matters by making decisions while listening on imperfect loudspeakers in less-than-ideal rooms [5]. A conscientious engineer can work to reduce this "confusion" and improve the final product by adopting the previously described practice of listening-room evaluation.

A properly treated acoustic environment is certainly preferred for monitoring a recording session but is rarely available when working on location. In particular, for projects in concert halls with or without an audience, the best location for the "control room" is generally somewhere backstage – in a dressing room or "green room" – the room where artists greet guests after a performance. Being situated close to the stage is a great time-saver when having to make last-minute adjustments to the microphone setup, and it is convenient for the artist when they wish to come and listen during session breaks or intermission. I can remember a certain conductor forgetting their tuxedo jacket after a playback in the Regie (control room) on the fifth floor at the Philharmonie in Berlin, and having to come all the way back for it while the audience waited patiently for the second half of the concert to begin!

2.5 Loudspeaker Listening

Loudspeakers and microphones are generally considered to be the two most important elements in the recording chain. Many experts believe that the loudspeaker is in fact *the most important* element, such that all decisions are influenced by the manner in which the recorded sound is presented to the decision maker. The question arises – which loudspeakers should you use? Should you bring your own, or use monitors provided by the studio (if an actual control room is available)? The answer is simple. Eliminate the variables as much as possible, to the extent this might be controlled:

- Bring the same loudspeakers to every session. This is a great way to gain some consistency in monitoring, when moving around from location to location. High-quality, full-range loudspeakers are the best option, but they are expensive and can be difficult and impractical to move around. Medium-sized active speakers (those common to recording studios in which the amplifier is built into the speaker cabinet) are a good option, since they are easy to transport.
- Choose from a selection of familiar loudspeakers provided by the studio. An engineer with several years of studio experience will become familiar with a common inventory of monitors available in studios worldwide. Once a favorite brand and model has been identified and "learned", it is worth requesting those same loudspeakers wherever you are working, if they are available. Following this practice can help greatly in terms of gaining consistency in the results, as it is a similar practice to bringing your own monitors.

With either of the above options, the interaction of the loudspeaker and listening room will still need to be evaluated, once again using a few well-known recordings for reference. It is dangerous to assume that having the same monitors at all times will guarantee consistency. The same set of loudspeakers will sound different in every room (just as their placement in the room will change their response); however, these variations can be observed and evaluated through careful listening. The lesson here is that by always reviewing the same "reference" recordings, an unfamiliar speaker setup can be learned and used effectively.

Is it best to use large or small loudspeakers? Full-range designs are generally best, especially for orchestra – but sometimes a medium-size loudspeaker will be less affected by room modes in the low bass range in a smaller-sized listening space, and therefore easier to work with. Many manufacturers of studio monitors offer controls for adjusting low- and high-frequency response. This can be helpful, but may also be confusing. These switches or pots should be set to a "flat" setting before listening, and then only used to help compensate for problems in the listening room after some evaluation period. Small monitors can be very helpful for balancing, and recommended over headphones for assessing solo balances, depth and reverberation. The low-frequency content of the recording should be carefully checked on good headphones in this case, as certain problems such as air-conditioning rumble or traffic noise may go unnoticed when using a small monitor in a listening room that is not perfectly quiet. Subwoofers may be used in conjunction with small monitors, although the setup should be carefully implemented and adjusted for even low-frequency response. Adjusting the physical placement of the subwoofer in the room will greatly affect how it performs in conjunction with the main loudspeakers.

2.6 Headphone Listening

There are certain pros and cons to headphone monitoring, and while headphone use should be limited as much as possible, they are an absolute necessity for recording sessions. High-quality headphones can be purchased for less than 10 percent of the cost of decent professional loudspeakers. Purchasing a good pair of headphones is an easy and affordable way to prepare for a professional career in "critical listening". Using headphones is a great way to monitor in a less-than-ideal listening environment, as the room can be almost completely removed from the equation. Headphones are also indispensable when working in a noisy space, or when complete isolation from the sound source is not possible (for example, literally "backstage" in a concert hall). In this case, "circumaural" or closed-back headphones will work best, offering the most isolation. They are also useful in checking for background noise from traffic, heating systems, low-level distortion from faulty equipment, or ticks caused by an improper word-clock setting.

The most negative attribute of headphone listening is that the audio image typically resides quite unnaturally inside the head, rather than out in front of the listener as is the case when listening to loudspeakers (see Figure 2.2). This phenomenon makes it very difficult to judge and properly

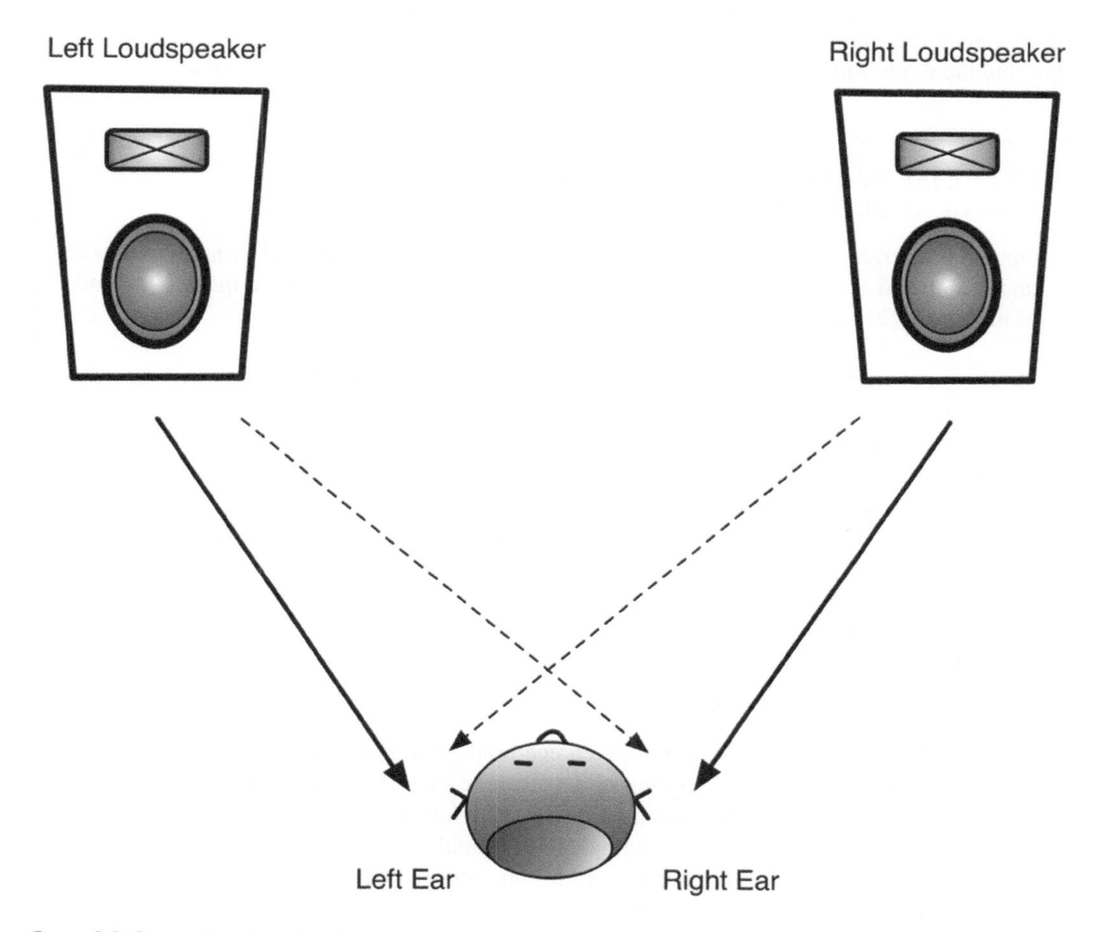

Figure 2.2 Stereo Loudspeaker Listening Showing Each Ear Receiving Both Signals

adjust the level of elements that appear in the center of the image, for instance a vocal or other soloist. Soundstage depth is impossible to assess properly, as is overall image width, since the ears receive totally separated left/right signals on headphones, whereas both ears receive information from both the left and right channels during loudspeaker listening.

Reverberation (natural or otherwise) tends to be more obvious on headphones, so the tendency is a recording and/or a mix that is rather dry when heard on loudspeakers. Low-level detail is also more apparent in headphones, which might potentially cause a microphone position that is too distant from the source. Headphones also come up short in providing the visceral experience inherent in loudspeaker listening regarding low-frequency information that is felt through the body rather than through the ear canal.

There are many different types of headphones, most typically either of an open- or closed-back design. Open-back headphones tend to sound more natural, and since they don't seal around the ear, the low-frequency response is more consistent with each wearing. They are also more comfortable to wear for long periods of time. Closed-back headphones offer greater isolation and generally deeper bass, but overall performance and consistency is greatly affected by how well they fit on the head. Closed-back headphones greatly inhibit communication in the control room, making it hard to communicate with other members of the recording team when everyone is wearing headphones. Imagine a producer asking an engineer for a balance change, and the engineer seemingly ignoring the request just because they are listening with closed-back headphones!

EXERCISE 2.3 COMPARING LOUDSPEAKER AND HEADPHONE LISTENING

Once a preferred set of loudspeakers and headphones has been selected, it is important to understand and acclimatize to both the similarities and differences between the two methods of monitoring sound.

1. Begin by listening to the loudspeakers, and using a few familiar sources of recorded tracks, set the volume to a comfortable and familiar level.
2. Changing over to the headphones, set the volume to a level that seems equal to the volume of the loudspeakers. If using open-back headphones, there will be some interaction with the loudspeakers unless they can be quickly set to dim or mute.
3. With the levels more-or-less the same, go back and forth between listening to the loudspeakers and the headphones, making note of all the differences.
4. Consider attributes or parameters such as frequency response, image width, transient response, amount of reverberation/spatial content, and so forth.
5. Take notes as seems appropriate.

With any luck, the differences between the choice loudspeakers and headphones can be learned and remembered, so that both monitoring options can be used interchangeably without a lengthy period of adjustment. In the worst case, looking for headphones that better match the loudspeakers may be a worthwhile investment. The engineer really shouldn't experience shock when switching to headphones after listening on loudspeakers for an extended period of time. This is especially relevant during the mixing process, as described in Chapter 24. Having multiple types and brands of headsets available is always an advantage, in order to have a selection of listening scenarios at hand.

2.7 Setting up the Control Room

There are several aspects to consider regarding the control room setup, and possibly the most important is to have a quiet space that is close to the stage. The location of the room should take into consideration the ease of cable runs to the stage, how the artist will get back and forth for playbacks and how the room can be securely locked between recording sessions over multiple days so that the equipment will be safe. Many halls with their own resident orchestra now have properly isolated control rooms, with varying amounts of acoustic treatment and available recording equipment. A room that is several floors away, which requires an elevator ride for access, may still be a good compromise if it is already fitted out with a console, loudspeakers and some decent acoustic treatment. The nuisance of having to travel a great distance to the stage is offset in this case by the potential increase in quality of the resulting recording. Permanently installed facilities tend to have fully tested cable runs in place, which can save time during the equipment setup and teardown.

In the case of a dressing room temporarily converted to a control room, it is important to consider the layout of the room for the best possible listening experience, but also to choose an orientation that makes the most sense rooms that people can enter and exit without disturbing the proceedings. Facing away from the door helps reduce distractions while working. Another important consideration is the cable run – it is best to try and avoid running lines across doorways, and through busy corridors, if at all possible. Even cables that are properly taped down can create a hazardous situation. Going over doorways is best, and always making sure the cabling looks tidy will help the operation have a professional look, even in a temporary installation. If the cabling doesn't fit under the closed door of the room, the cables will need to be easily disconnected at that point, so that the door can be closed and locked when the crew is not around.

Once all the equipment is up and running, don't forget to listen again. This new room, like all control rooms, must be "learned" in the same way – by playing some familiar recordings. What do the loudspeakers sound like when they are stuffed into a dressing room backstage at a concert hall? A rapid assessment is recommended, so that a conclusion is reached before the brain has a chance to adapt to the anomalies of the room – otherwise the acoustic deficiencies of the listening room may adversely affect microphone placement and balance. More often than not, major issues will have to be endured rather than corrected in a listening room that is thrown together in a hurry.

Low frequency is particularly hard to modify without making major changes to the room. Simple adjustments can be made that will help – moving the loudspeakers closer to the wall will increase the low-frequency response, and a dark or dead room can be improved by decreasing the distance between the speakers and the listener. A fully carpeted room can be made more live by placing a few sheets of plywood on the floor near the loudspeakers. These suggestions will make a small change, and the listening setup will most likely remain compromised, to a certain extent. It is impossible to think that a makeshift room will be much improved by hanging some blankets or putting down a carpet. In fact, a room with low-frequency resonances is usually made worse when packing blankets are added, as these tend to soak up energy in higher frequencies, which will then magnify the original issue.

It is a good practice to choose a comfortable listening level and to stick with it. Just as your "well-known" recordings act as a reference, listening at a set volume will help make informed decisions over the course of the hours and days of a project. Make a note of the volume level if a readout is available or make a mark in pencil next to the line on the volume control on the console or interface being used, so that it is always possible to get back to normal if the volume is lowered during a break, or raised to check for noise or low-level details. The same practice should be followed for headphone volume, so that most listening is done at a set reference level. There is a natural tendency to want to sneak up the volume when the ears become tired over long periods of listening – this should be avoided, and in fact the volume should really be lowered if listening fatigue is in play.

2.8 Listening "Live" to the Microphones

Opening up the microphones and assessing the result will be discussed in greater detail in Part II, but it should be mentioned here how this practice falls in line with previous Sections 2.3 and 2.4. Again, make quick decisions here, in order to avoid getting used to the sound. The aural memory can be erased and "refreshed" by muting the microphone signals and waiting a few seconds, switching to headphones or quickly listening to something else. Be careful, however, not to judge the sound or placement of the microphones prematurely when listening to an orchestra that is not warmed up. Even in a professional orchestra in which the individual members arrive early to prepare, there is a strong tendency for the brass to overplay in the first few minutes of the session, and for the string sound to begin to open up and sound fuller after the first 10 or 15 minutes.

2.9 Chapter Summary

- Critical listening is the most important skill for an audio engineer to develop.
- Training the brain is the key to improving the acuity of one's hearing.
- Bring reference audio examples with you wherever you go.
- Loudspeakers, along with microphones, are the most important tools used in audio recording.
- Use a combination of headphone and loudspeaker monitoring for best results.

References

1. Wolfe, J.M. et al. (2006). *Sensation and Perception*. Baltimore, MD: Sinauer Assoc, pp. 76–154.
2. Eargle, J. (1980). *Sound Recording* (2nd Ed.). New York: Van Nostrand Reinhold.
3. Corey, J. (2013). *Audio Production and Critical Listening: Technical Ear Training*. Burlington, MA: Focal Press.
4. EBU Website (2016). Under Publications, "EBU Tech 3286: Assessment Methods for the Quality of Sound Material – Music", https://tech.ebu.ch/publications/tech3286, Aug. 1997.
5. Toole, F. (2008). *Sound Reproduction*. Burlington, MA: Focal Press.

Chapter 3

Understanding Microphones

This chapter reviews the fundamentals of microphone characteristics, including a discussion of the functionality of each type of microphone and its use in standard stereo microphone techniques. References are included to various sources that provide more technical information for those readers who have an interest in exploring the intricacies of microphone design. Newcomers to audio recording will be introduced to the various characteristics of each microphone type, and the most common techniques for two-microphone stereo recording. Some general traits and "best uses" will be addressed, along with a short section on microphone preamplifiers. Recording engineers need to prioritize the use of their available microphone inventory – saving the best microphones for the most important applications within the recording process – whether they have 4 microphones or 24.

3.1 Introduction to Microphones

It is important to understand that microphones can be classified in several manners, and therefore described using various attributes. Transducer design, directional sensitivity and resulting polar pattern, as well as the type of power supply, must all be noted. It is possible to have an omnidirectional dynamic microphone, or a condenser cardioid that may or may not use a tube in the power supply design. More detailed information can be found in [1–3].

3.2 Transducer Type

By definition, a transducer is any device that changes energy from one state into another. In audio, microphones are used to transfer acoustic energy into electrical energy, and loudspeakers perform the same function, except they do so in the reverse direction, via a mechanical apparatus. Microphones perform this function through the use of a diaphragm, and the two most common designs for recording music are dynamic and condenser.

A *dynamic* microphone generally refers to a design in which a moving coil is suspended in a magnetic field and is attached to a diaphragm that is caused to vibrate by changes in pressure in a sound field. An example is shown in Figure 3.1.

Simply put, the movement of the diaphragm generates an electrical signal in the coil as it moves back and forth within the magnetic field (see Figure 3.2).

DOI: 10.4324/9781003319429-4

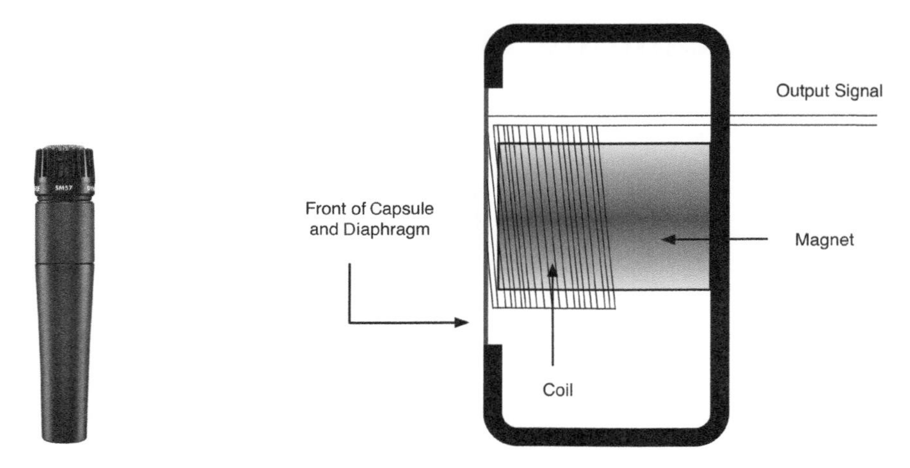

Figure 3.1 Shure SM57 Dynamic Microphone

Credit: Shure Inc. (used with permission)

Figure 3.2 Diagram of a Dynamic Moving Coil

Ribbon microphones are also dynamic microphones, but they use a thin corrugated piece of metal suspended in a magnetic field rather than a coil of thin wire: an example is shown in Figure 3.3.

Due to its small size and mass, the ribbon generates very little signal, so the output level is very low. A transformer is needed on the output to compensate for a very low impedance (see Figure 3.4). Some ribbon microphones are now available with active preamps, which raise the output level and correct the impedance for optimized use with any microphone preamp. Ribbon microphones are sometimes referred to as "velocity" microphones, as the rate at which the ribbon moves is directly proportional to the pressure difference on each side of the ribbon.

> While the two types of dynamic microphones are most accurately known as dynamic moving coil and dynamic ribbon, each type is more commonly referred to as simply "dynamic" and "ribbon".

Condenser microphones are less commonly known as capacitor microphones, so named because the diaphragm of the microphone is essentially one plate of a capacitor (or condenser). An example can be seen in Figure 3.5. Changes in sound pressure affect the distance between the two plates, which in turn varies the voltage according to the capacitance (see Figure 3.6). As one can imagine, the displacement of the capacitor plate is so small that the resulting variation in voltage requires amplifying before it leaves the microphone. As such, condenser microphones have amplifier circuits built in, and these require a voltage to power the "on-board" electronics, whether they are of a tube or transistor design. Some microphones of a single-capsule solid-state condenser design utilize an electret capacitor, so that the diaphragm is permanently polarized. This differs from regular capacitors, where a voltage is applied to polarize the two plates. Tube-driven microphones have an external power supply with a multi-pin cable that connects to the microphone. In this case, the audio signal runs separately from the voltage that powers the electronics and polarizes the capsule.

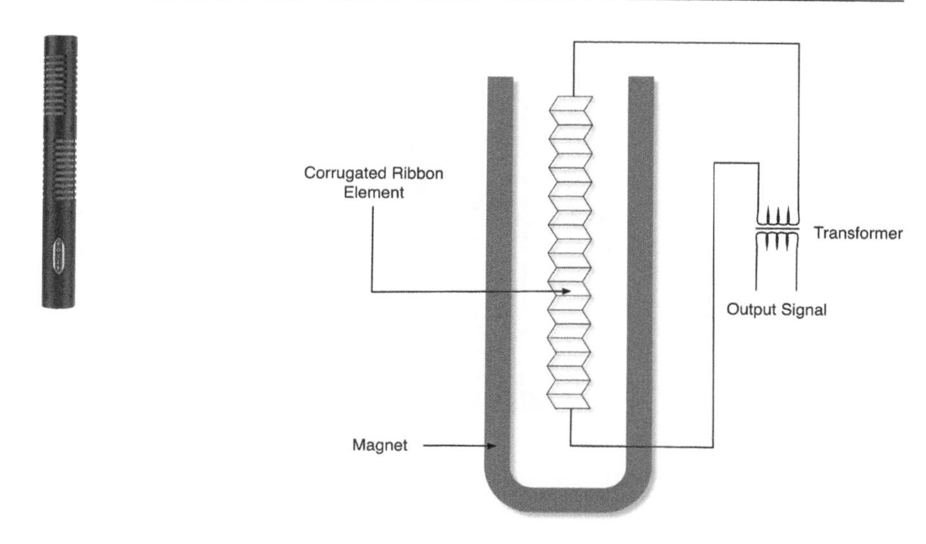

Figure 3.3 Royer Labs SF-12 Stereo Ribbon Microphone

Credit: Royer Labs, USA (used with permission)

Figure 3.4 Diagram of a Ribbon Transducer

Figure 3.5 DPA 4011 Condenser Microphone

Credit: DPA Microphones (used with permission)

Phantom power: Condenser microphones are designed with a preamplifier located inside the body of the microphone. In this case, an external voltage is needed to power the electronics and occasionally to polarize the capsule,, and this voltage is "carried" on the same cable that transmits the audio signal. This is called phantom power since it provides a voltage to the microphone's preamp while remaining "transparent" to the audio signal, and it is filtered or removed at the audio input of the microphone preamplifier. A simple explanation is that it is a "steady-state" or DC voltage upon which the audio signal is transmitted.

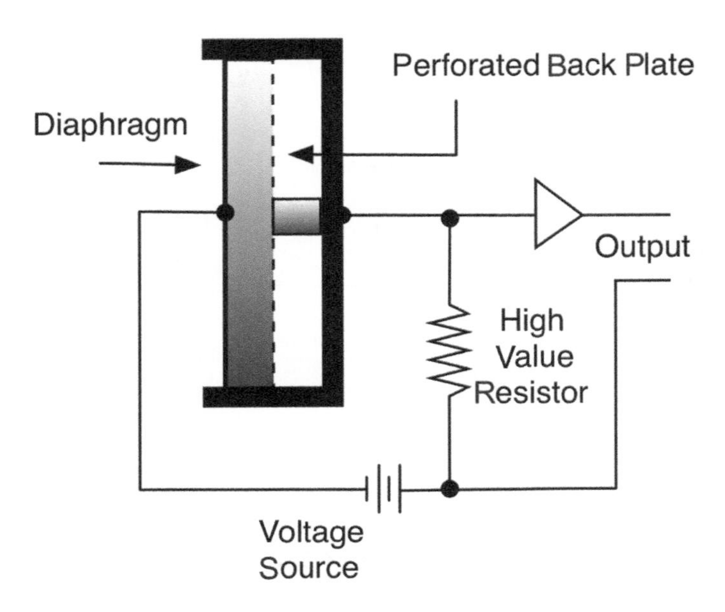

Figure 3.6 Diagram of a Condenser Transducer

These three designs, dynamic, ribbon and condenser, have pros and cons that should be noted, so that each one can be applied appropriately based on its particular characteristics (see Table 3.1):

Sensitivity: Normally, in classical music recording, for the principal microphone pickup one will want to use microphones that have the greatest sensitivity possible. In order to get a decent overall capture of the ensemble, the microphones will need to be a certain distance from the source (more on this in Section 6.1, Main Microphone Systems). Microphones with reduced sensitivity will come up short here, and as such, condenser microphones are the obvious choice.

Placement to source: I remember hearing a comparison of ribbons with condensers based on a recording of a triangle at very close range. The ribbon sounded very smooth and silvery, whereas

Table 3.1 Comparison of Transducer Designs

	Dynamic Moving Coil	Ribbon	Condenser
Sensitivity	quite low at high frequencies	very sensitive in high freq. range	high
Placement to Source	optimized for close placement	best at close range	can be used at varying distances
Transient Response	slow	quite fast	very fast
Ruggedness	very rugged	very fragile	somewhat delicate
Output Transformer	usually needed	always needed	not always required
Resonant Frequency	large peak in mid to upper-mid range	in very low freq. range	in high freq. range, subtle peak
Frequency Response	erratic, with large peak at resonant frequency	weak in LF, smooth but rolling off in HF	normally quite flat
Cost	inexpensive	mostly expensive	affordable to very expensive

the condenser sounded as if the triangle beater (or the triangle itself!) had been used to strike the capsule of the microphone. Of course, the comparison was unfair, as the condenser microphone's inherently superior sensitivity meant it had no business being placed so close to the triangle. Again, it would seem that condensers are a natural choice for classical music recording, but the other microphone types have their place as well. There are situations where close proximity to sources is necessary in order to have control over balance within the sections of the orchestra, and at such times, ribbons in particular can be a very useful tool.

Transient response: When considering this attribute, one might assume that the microphone that reacts the "fastest" will give the most natural and appropriate result. There are times, however, when the source signal will want to be "padded" or relaxed, and a "slower" microphone may be just the ticket. Certain percussion and brass instruments may benefit from a microphone with a slower response time when recording at close range.

Ruggedness: Of course, most microphones are inherently fragile, as the diaphragm will not perform correctly (or at all) if it is damaged. Dynamics can generally take a beating (literally), whereas ribbons require great care in handling and exposure to sudden bursts of air movement. Most ribbons come with a fabric cover that can be kept over the diaphragm until the beginning of the recording session. Condensers with solid-state electronics are fairly sturdy but should be treated very carefully – especially when connecting and disconnecting. Condensers are also more sensitive to humidity than dynamic microphones. Tube microphones should be handled with extra care because of the delicate electronics and the usually fragile multi-pin cable, which is attached using a threaded connector that can be easily damaged or stripped.

Transformer: This is mentioned here because many companies offer transformerless designs of condenser microphones. Transformerless designs are advertised as sounding more "open" and transparent, and as having a lower noise floor, although a transformer can have a positive effect on a somewhat "plain"-sounding source that might benefit from having some nonlinearity or "color" provided by the microphone's electronics. This effect might be considered less beneficial in classical music recording.

Resonant frequency: An important characteristic to consider when choosing a microphone for a particular application, resonant frequency has a great influence on timbral characteristic. Some manufacturers claim that the resonance is in the low-frequency range, when in fact it is much higher up. A low-frequency boost may be apparent, but this is caused by design factors other than resonant frequency.

Frequency response: This characteristic is probably the most critically observed. Generally, a microphone with flat response will provide the most faithful capture of the source, although certain microphones with an "erratic" response may be used where a specific color is desired. Microphone manufacturers frequently alter frequency response to create designs that are specialized for certain uses, such as "kick drum" microphones in pop music recording.

Cost: For engineers at the start of their careers, this factor will be very important. Good-quality ribbon microphones tend to be quite expensive, whereas entry-level condensers are quite affordable. Many very good dynamic microphones are available at a reasonable cost. The most expensive microphones will be the large-diaphragm multi-pattern condenser microphones with a tube power supply.

Based on a comparison of these attributes, it is apparent that condenser microphones are the "mainstay" of classical music recording. The cost factor is worth discussing in this respect, and by default the best microphones are generally the most expensive, but with condensers it is possible to put together an affordable start-up kit that can be upgraded over time. The advantage is that the characteristics of the condenser design can be investigated using an inexpensive model. This way, recording techniques can be learned and refined before the purchase of high-quality microphones becomes a practical venture.

3.3 Directional Sensitivity

In the previous section on transducer type it is understood that microphones are designed to respond to changes in sound pressure. Another important part of the design is that of directionality, or variations in sensitivity around the diaphragm (front, sides and back). Each design can therefore be described by its resulting "polar pattern", which can be easily viewed and understood in a graph format. For simplicity, polar patterns are shown in two dimensions, although the patterns exist in three dimensions. For example, when discussing the sensitivity at "side" of the capsule or 90° off-axis to the face of the diaphragm, this includes positions to the sides, above and below. An omnidirectional polar pattern is actually spherical in shape, although it is shown as a circle in Figure 3.7.

Omnidirectional "omni" or pressure microphones respond to pressure almost equally from all angles around the diaphragm, except in very high frequencies.

In theory, one might assume that the orientation of the placement (the direction in which the microphone is pointing) is unimportant. Since the microphone becomes somewhat directional in high frequencies, it is necessary to carefully consider the orientation of the capsule in relation to the source. Through their simplicity of design, omnis have very even frequency response extending to both the low and high range. Since this design offers very little rejection of sound coming from behind the microphone, its functionality is somewhat limited. The only real option for controlling balances is through the physical placement in the recording room. That being said,

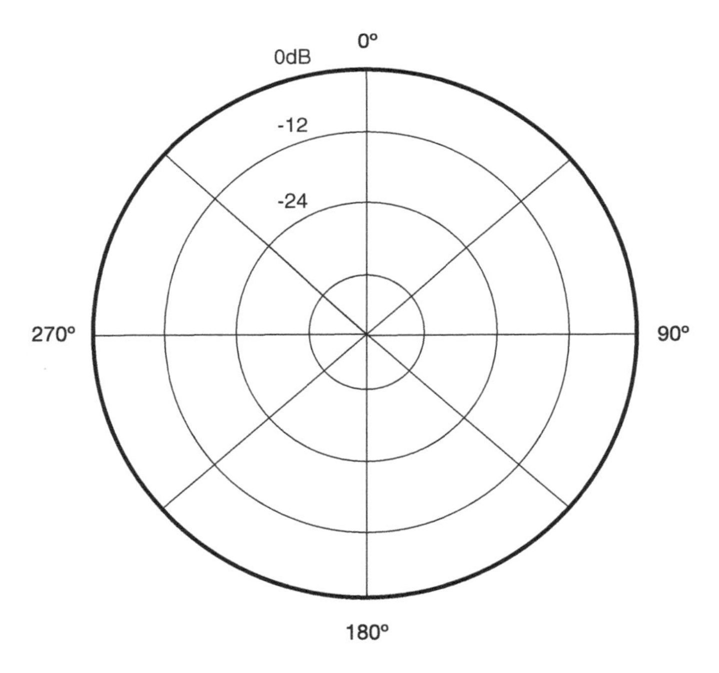

Figure 3.7 Theoretical Polar Pattern of an Omnidirectional Microphone

using omnidirectional microphones in a good room can be a powerful tool for achieving a natural, highly refined overall sound, which normally requires very little processing. Omnidirectional microphones are much less susceptible to wind noise than cardioids, which makes them more suitable in outdoor concert recording, at least for a main pickup system.

Bidirectional or "figure-8" microphones work on the principal of pressure gradient, which describes the direction of the sound wave in relation to the orientation of the diaphragm, or the difference in pressure between the front and back of the diaphragm (see Figure 3.8).

Pressure directly in front of the microphone will be observed as 180° out of phase with sound pressure that is present on the back of the diaphragm. Sound "arriving" from the sides reaches both the front and back at the same time, so that signal is canceled. The resulting polar pattern resembles the figure "8", hence the name figure of 8, or figure 8 microphone.

Directional or "cardioid" microphones using a single capsule utilize a system of ports or vents to allow sound pressure from behind the microphone to reach the front and back of the diaphragm in opposing polarity, thereby partially canceling the signal from the sides, and almost completely canceling the signal arriving behind the capsule (see Figure 3.9). In the case of a single-capsule microphone, this is accomplished through the addition of tiny holes in the back plate of the capacitor that allow for the presence of sound pressure on the backside of the diaphragm. The term cardioid is used because of the resulting heart-shaped polar pattern. A hypercardioid polar pattern is a more directional version of cardioid, realized by a more complex set of ports in the microphone body just behind the capsule. Shotgun microphones are simply a more extreme version of the hypercardioid, requiring a much longer body to accommodate a longer series of ports. (An example of the hypercardioid polar pattern is shown in Figure 3.12.)

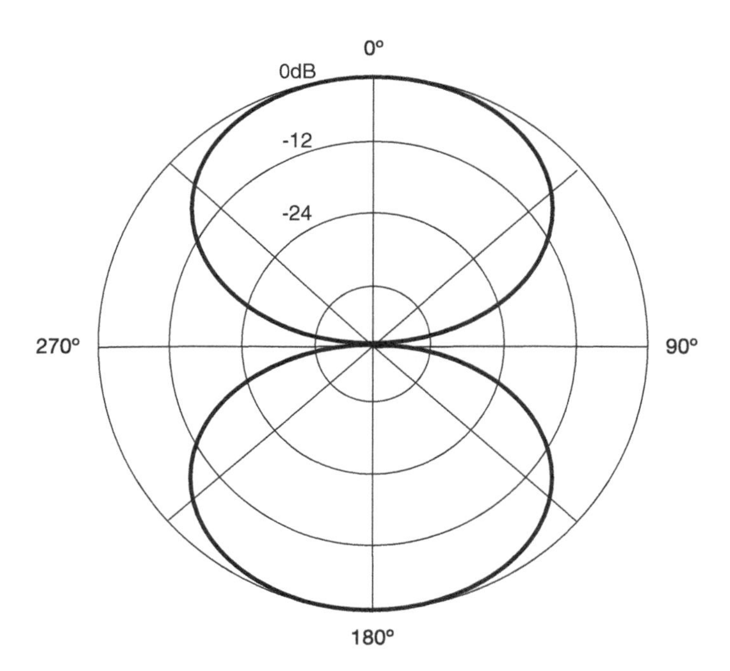

Figure 3.8 Theoretical Polar Pattern of a Bidirectional Microphone

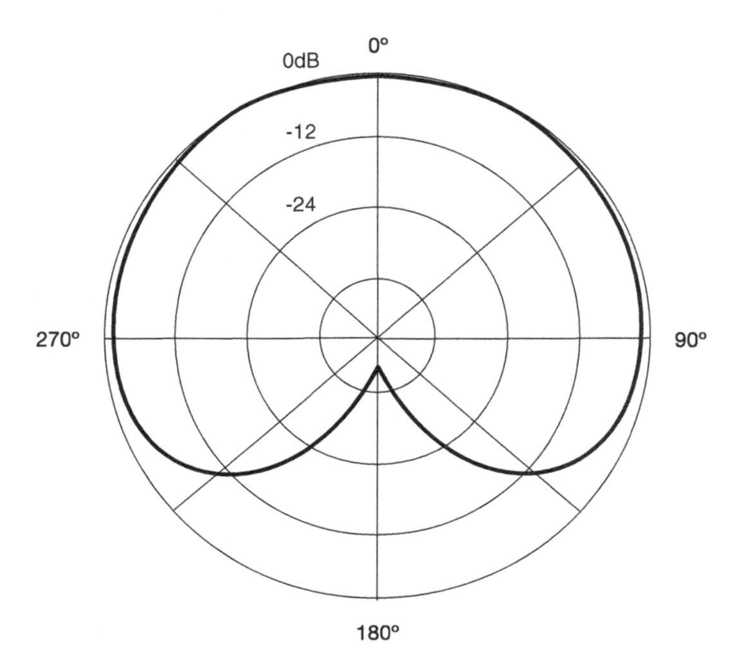

Figure 3.9 Theoretical Polar Pattern of a Cardioid Microphone

Cardioids are actually a combination of pressure and pressure gradient design, as they respond more like omnis at low frequencies, while they become more directional toward high frequencies due to the gradient effect of the canceling ports.

There are many combinations of transducer type and polar pattern – an omni can be of a dynamic or condenser design, and a figure 8 might be condenser (solid state or tube), or ribbon, so it is important to know of all the characteristics and how each influences the resulting sound capture.

Multiple polar pattern microphones come in two versions. The simplest design is mechanically switchable between cardioid and omni by simply closing off the ports or vents behind the diaphragm. In fact, any ported cardioid can be theoretically "converted" into an omni by placing a piece of adhesive tape over the ports – although the resulting frequency response may not be very even, depending on the microphone design. The more common multiple-pattern microphone incorporates a dual-capsule design of back-to-back cardioids, in which the outputs of the two capsules are electrically summed using a combination of polarities and gains to create different polar patterns at the output. An example is shown in Figure 3.10.

Originally developed by Braunmühl and Weber in the 1930s, the simplest version provides for omnidirectional, cardioid and bidirectional patterns. Cardioid pattern is achieved through a single output of the front-facing capsule, omni is the sum of the two capsule outputs, and bidirectional or figure-8 pattern is the result of the two capsules summed in opposite polarity (front is positive, rear is negative; see Figure 3.11).

More versatile designs offer two additional polar patterns – subcardioid (or wide cardioid), which is a combination of omni and cardioid, and hypercardioid (between cardioid and figure 8). Subcardioid sums the two capsules in positive polarity, with reduced gain from the rear capsule, and hyper applies the reduced gain of the rear capsule in reverse polarity (see Figure 3.12).

Figure 3.10 AKG 414 XLS Multi-pattern Condenser Microphone

Credit: AKG Microphones (used with permission)

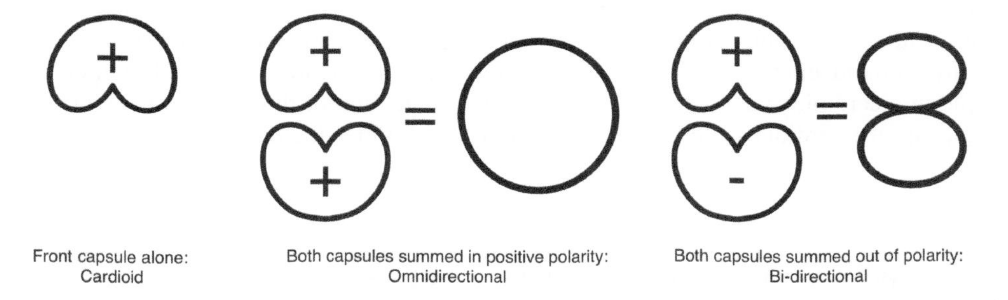

Figure 3.11 Diagram of Polar Patterns Produced by Combining Two Capsules

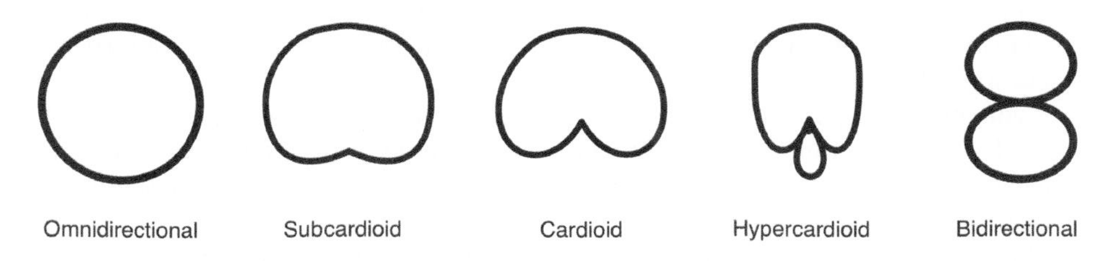

Figure 3.12 Diagram of Five Polar Patterns Produced by Summing the Two Capsules with Varying Gains

A five-position switch is used, or in the case of a variable control on a tube power supply, these patterns are available by setting the variable control between the three "standard" patterns.

In this design, either phantom power or an external power supply is used to polarize the capsules, as it needs to be changed depending on the desired polar pattern. In the case of tube microphone designs it is common for the polar pattern to be selected on the power supply unit rather than the body of the microphone, and as is typical of early designs the control is variable between patterns, which offers even more versatility of polar pattern in comparison to a stepped switch – for instance "supercardioid", which is somewhere between a cardioid and a hypercardioid. An interesting side note here – a change in polar pattern in this design will also affect the tonal characteristic of a multi-pattern microphone. For example, switching from cardioid to subcardioid normally results in a darker frequency response combined with the wider polar pattern.

It should be stated that a multi-pattern microphone does not replace the need for single-capsule designs with dedicated polar patterns. In fact, a dedicated omnidirectional microphone with a single capsule generally provides the most even response and the most perfect "omnidirectional" characteristics, except for a narrowing in directional sensitivity into the higher frequencies. A dual-capsule design will exhibit a narrowing on both the front and back plates at higher frequencies. In cardioid mode, the dual-capsule design tends to have a much wider pattern, where the single-capsule cardioid rejects more sound pressure from the sides and back because of the efficient porting system. Interesting to note, however, that in close proximity to a source the dual-capsule design remains more "cardioid" in low frequencies, and a single-capsule ported design becomes bidirectional around 100 Hz and lower.

Reverberation radius, direct-to-reverberant ratio: It is important to bear in mind that each polar pattern exhibits a different reverberation radius, or point at which the direct-to-reverberant ratio will match the other polar patterns. The omnidirectional polar pattern is used as a baseline measurement with reverb radius of "1", and the other patterns are compared to that reference. Reverb radius for cardioid and figure 8 is 1.7, supercardioid is 1.9, hypercardioid 2.1 and shotgun 2.2. For example, if a cardioid and an omni were to have the same ratio of direct-to-reverberant sound in their pickup, the cardioid would have to be placed at 1.7 times the distance to the source of the omni.

EXERCISE 3.1 COMPARING POLAR PATTERNS

This is a simple yet informative test to help understand and compare polar patterns:

1. Set up a microphone, and have someone walk around it while talking normally, trying to keep the same distance from the capsule at all times, and announcing their position relative to the diaphragm – 0° when directly in front, or "on-axis", 90°, 180° (behind), and so forth.
2. If you are alone, you can record your own "performance" and then play it back afterward. The effect can also be experienced in real time by wearing headphones while speaking and moving around the microphone – as long as care is taken not to trip on the cables while moving around!
3. Alternatively, the microphone might be rotated in place while a source signal or a person speaking remains in a fixed position, as long as the rotation of the microphone can be done quietly, using a turntable or similar support.

3.4 Stereo Microphone Techniques

While Chapter 6 will engage in a more in-depth discussion of main microphone systems for orchestra, this section will provide an overview and explanation of some standard stereo microphone techniques. The simplest (and arguably the best) way to make a stereo recording is to use a pair of microphones with matching characteristics and apply the outputs of each microphone to two inputs (left and right) of a recording device.

> One of the "ground rules" of stereo recording, according to John Woram, is: "*NEVER use more than two microphones*" (emphasis in original) [4]. While this statement is quite correct in theory, it is mostly impractical in commercial orchestral recording. The rest of this book will focus on how to properly break this rule, while remaining respectful of the sentiment that lies therein.

As soon as a third microphone is introduced into the mix, both positive and negative results will be observed, such as cancellation, distorted image or perspective, or a change in overall evenness in frequency response. For example, introducing a signal from a microphone that is placed near an instrument in the back of the stage (e.g., trumpet) will help to bring it forward in the image, thereby collapsing the natural depth of the recording or distorting the perspective. This might be considered a good thing if the trumpet is too distant sounding, but it may also serve to make the trumpet abnormally close or loud if the signal is overused.

The objective is to combine various microphone signals using a methodology that benefits from the positive side effects while minimizing the negative artifacts. The same might be said of adding signal from the bass section, which might help to add low-frequency information but may in fact create an exaggerated low-frequency response if overdone (more on this subject in Chapter 24, Mixing). The best place to begin is in reviewing, evaluating and deciding upon the use of a stereo system for the overall capture of an ensemble or sound source, using a conventional technique or a variation thereof. The following sections describe the most common stereo recording techniques in use, along with their positive and negative attributes. These techniques can be directly compared on the companion website (https://routledgetextbooks.com/textbooks/9781138854543), via a solo piano excerpt using AB, ORTF, XY, MS and Blumlein techniques.

XY (crossed cardioids): This is a coincident technique, meaning the capsules of the two microphones virtually share the same space. Typically, the capsules are positioned one above the other, so that there is no difference in the time of arrival of signals reaching the system from left to right (see Figure 3.13). In XY, the microphones are usually cardioids and are most commonly positioned at an angle of incidence of 90°.

The stereo image is achieved based on the intensity differences of the signals reaching the two microphones. Sound coming from the left side will have a greater pressure on the left microphone, so it will be "heard" on the left side of the resulting recording – this principle is known as Interaural Intensity Difference, or IID.

Overall image is very stable, as is phantom center. XY is mono compatible, as there is no cancellation caused by phase (time) differences when the left and right channels are combined. This technique is best used for small sound sources or perhaps soloists, and it should be noted that center signals are off-axis to both capsules, so the system requires the use of high-quality microphones with an even frequency response on the sides of the capsule. More distant sources will pull

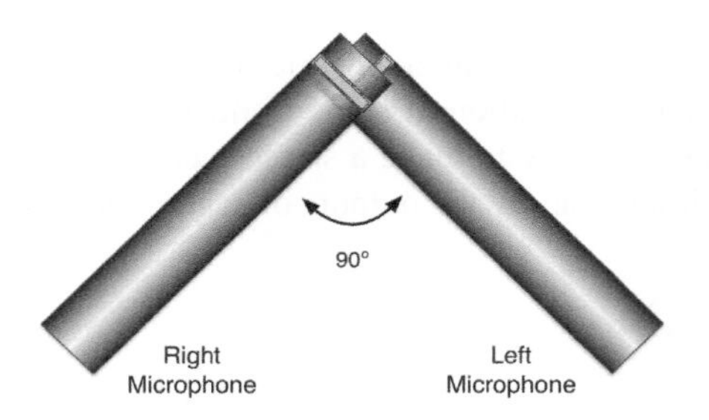

Figure 3.13 Diagram and Photo of XY Technique

Credit: Photo: Jack Kelly

toward the middle, and the system in general lacks spaciousness. A build-up of low-frequency information may be observed in the center image using XY technique, as cardioids operate as pressure transducers (rather than pressure gradient) in the low-frequency range.

Blumlein (crossed figure 8s): This is a coincident technique similar to XY, in which a pair of bidirectional microphones are placed at 90° and one above the other (see Figure 3.14).

Again, the advantages are stable image and mono compatibility. Also, the cross-channeled, reverse-polarity back-plate information can add density to the reverberant field. In other words, information coming from the rear of the left-hand side of the system is captured by the reverse-polarity back plate of the right channel microphone. As such, careful attention must be given to information arriving from the rear and sides of the system, and sidewall reflections can be problematic. As with the other coincident techniques, Blumlein does not offer great width in its coverage and is therefore best used for recording single instruments or small ensembles.

M-S (mid-sides or mono-stereo, aka sum and difference). This is also a coincident system where two microphones are used to generate an adjustable-width stereo pickup by capturing a mono signal separately from the diffuse/indirect sound. The "M" component can be recorded using any polar pattern directly facing the source (omni, cardioid or figure 8), whereas the "S" component is

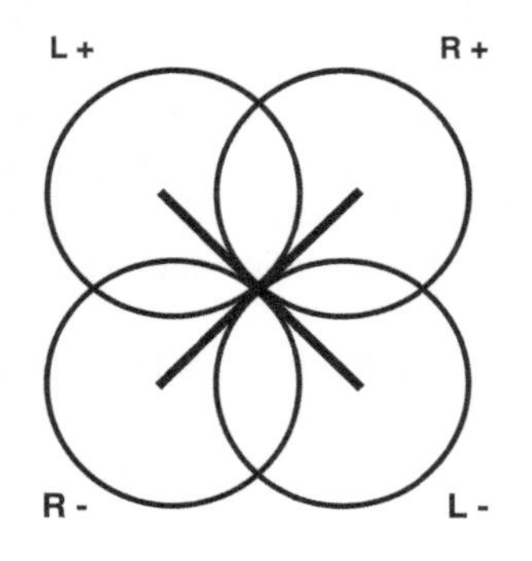

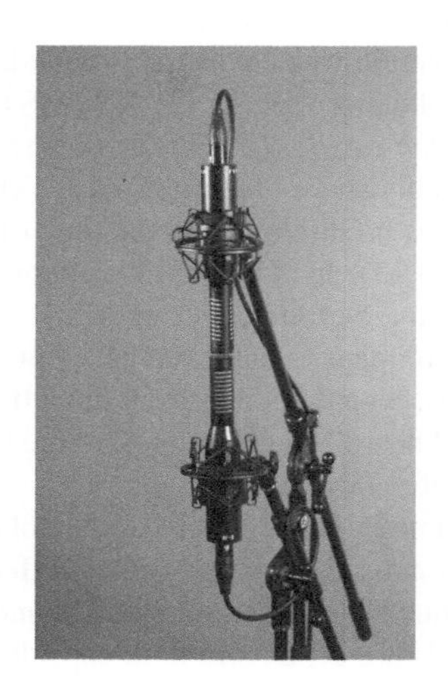

Figure 3.14 Diagram and Photo of Blumlein Technique

Credit: Photo: Jack Kelly

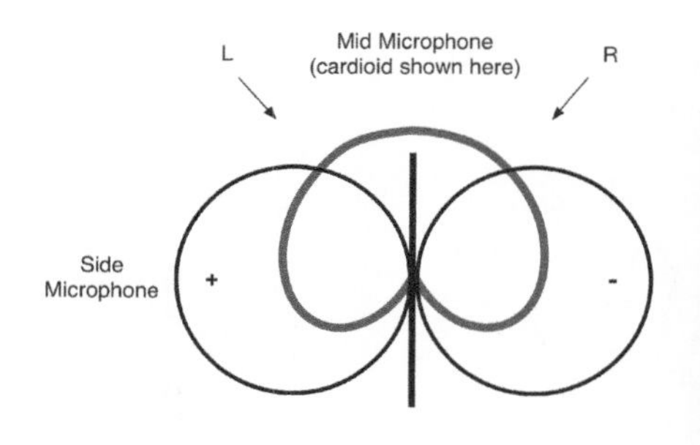

Figure 3.15 Diagram and Photo Showing M-S Technique

Credit: Photo: Jack Kelly

captured using a figure 8 placed at 90° to the source and directly under or above the M microphone, with the positive side of the capsule facing to the left (see Figure 3.15).

The amount of mono vs. stereo information can be controlled from the listening room, and if the two microphones are recorded separately, the width can be adjusted after the fact as well. This system is also highly mono compatible and free of off-axis coloration. The mid microphone is applied equally to both left and right channels (panned center), while the signal from the figure-8 microphone is split and applied equally to both left and right, but with right channel in reverse polarity (see Figure 3.16).

Changes in the balance of the mid (L+R) and sides (L-R) components affects the overall width of the image. Care should be taken in placing the microphone so that the front side (positive) of the capsule is facing the left side of the source. While the system is quite flexible and allows for refining the perspective after the fact, it requires additional work to set up. Also, narrow rooms and parallel walls can be problematic with this technique.

AB or *spaced pair*. This technique relies mainly on time of arrival to create the stereo image, in which Interaural Time Difference or ITD is responsible for where sources appear in the lateral soundstage (see Figure 3.17). Signals coming from the left side simply arrive at the left microphone ahead of the right, so that the sound source is heard on the left side of the recording and playback audio.

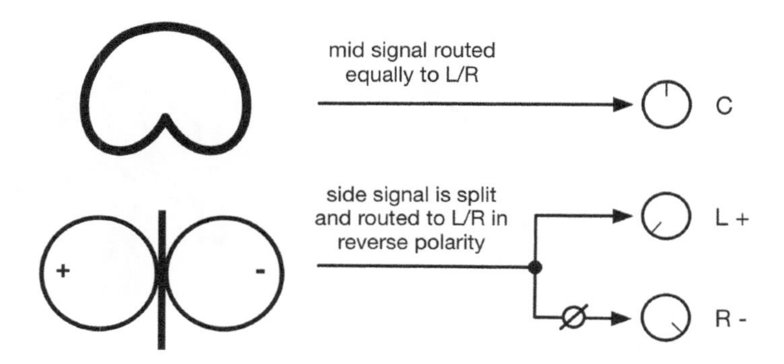

Figure 3.16 Diagram of M-S Monitoring

Figure 3.17 Diagram and Photo of AB Technique

Credit: Photo: Jack Kelly

When a spaced pair of omnidirectional microphones is used, there is very little difference in intensity between sources arriving from the left and right, except for extremely wide stage placements. Different polar patterns can be used, typically from omni to cardioid, and the spacing of the system depends on the size of the source or ensemble and the desired width in image. AB tends to have an open, spacious quality with good warmth (especially when using omnis), and off-axis coloration is less of an issue in smaller configurations (soloist for instance), as both microphones are facing directly toward the source. The drawbacks are poor mono compatibility, exaggerated depth of sound field and a potentially unstable image and/or reduced center information (hole in the middle) if the microphone spacing is too wide.

ORTF: This is a "near-coincident" system developed by the Office de Radiodiffusion Télévision Française, now operating as Radio France, and works on the combined principles of IID and ITD – using both intensity and time of arrival to properly localize sources within the stereo image. Specifically, the system consists of two cardioid microphones with the capsules spaced at 17 cm and at an angle of 110° (see Figure 3.18).

The design was optimized for covering large ensembles with good sense of depth and a decent – although reduced – mono compatibility, which is important for broadcast. Close placement to the source may be somewhat unstable.

NOS. A variation on ORTF, used by the Nederlandse Omroep Stichting, or Dutch Broadcast Foundation. Cardioids are placed 30 cm apart, at an angle of 90°, such that the system relies more on ITD than IID (see Figure 3.19).

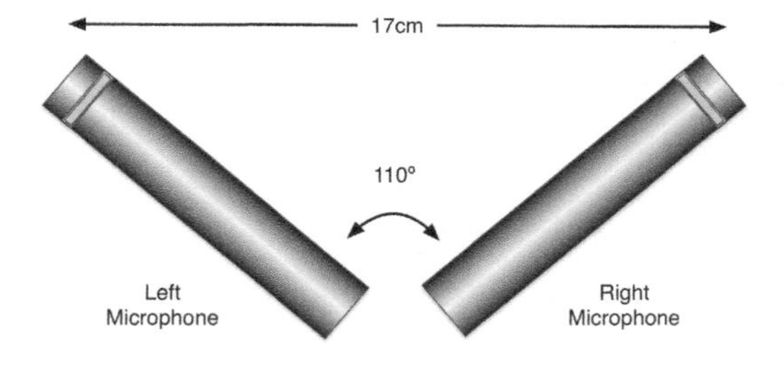

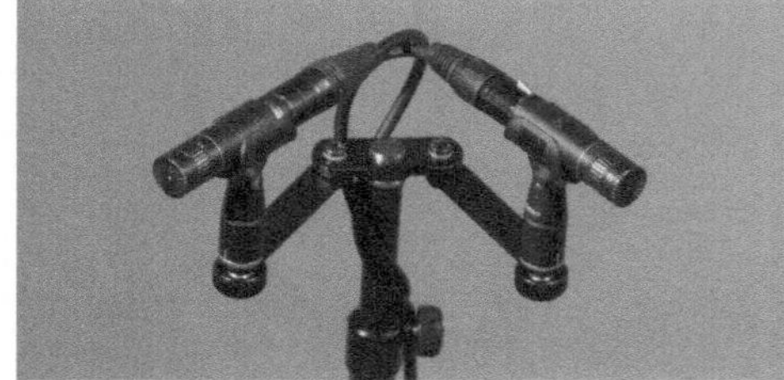

Figure 3.18 Diagram and Photo of ORTF Technique

Credit: Photo: Jack Kelly

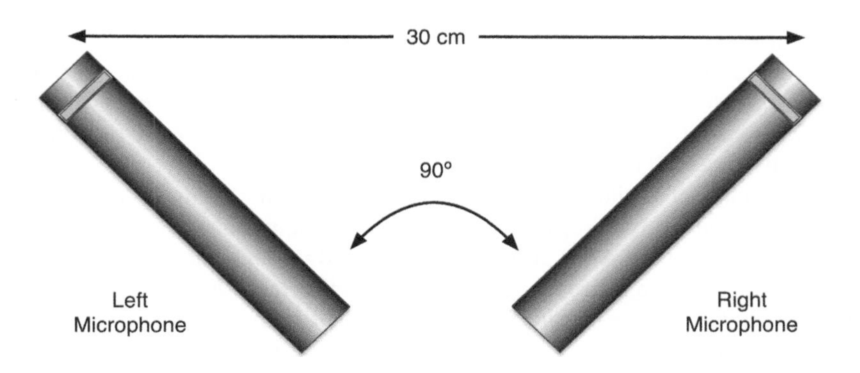

Figure 3.19 Diagram of NOS Technique

Table 3.2 Comparison of Stereo Microphone Techniques

	X-Y	Blumlein	M-S	AB	ORTF/NOS
Source Size/Coverage	small/single	small	small	small to very large	best for large
Mono Compatibility	high	excellent	high	moderate to poor	very good
Spaciousness	poor	minimal	moderate to poor	excellent	moderate
Low Frequency Response	poor	poor	varying	excellent	moderate
Depth of Image	poor	moderate	moderate	exaggerated	very good
Stability of Image	excellent	excellent (on axis)	excellent	moderate	very good
Ease of Implementation	moderate	poor	poor	excellent	moderate

The result is that it is slightly more spacious and less mono compatible than ORTF – otherwise the two systems are quite similar. The slight difference can be attributed to a subjective preference of one group of engineers as compared to that of another. A comparison of stereo microphone systems and their pros and cons is provided in Table 3.2.

When it comes to purchasing microphones on any budget, it is worth making sure that an appropriate amount of funds is allocated to the purchase of the main pair, as in most cases, these two microphones will be responsible for providing the basic characteristics inherent in the audio capture.

3.5 Microphone Preamplifiers

It is worth mentioning that some of the issues that result from microphone design and characteristics are relevant to the microphone preamplifiers as well. Preamps, which boost the output of the microphone to a recordable standard "line level" signal, come in just as many different designs – solid state, transformer and tube implementations. The use of an interface or recording device that has "on-board" preamps should still involve a proper evaluation of the preamp's characteristics. Regarding Toole's "Circle of Confusion" (Section 2.4), the more the engineer is aware of the different attributes of each link in the audio chain, the better the outcome of the recording. Knowing the available tools very well is key to a successful recording session.

Comparing preamps of similar design can be very educational. Various preamps which manufacturers will guarantee are "perfectly flat" in frequency response may sound very different when auditioned against each other. After proper evaluation and through careful listening sessions, an engineer can properly pair up microphones with preamps that result in a satisfactory combination. For example, a very fast condenser microphone may benefit from being paired with a transformer or tube preamp with a "smoother" or slower transient response, in certain circumstances. Certain practical observations may be noted as well, such as a brand of preamp or a design that works well in terms of its ability to react to intense amplitude modulation and the resulting input voltage, which are common to some situations such as vocal recording. As with microphones, the preamps to be used with the main microphones should be the best quality available, and preamps of a lower caliber should be reserved for less important support microphones – those signals that will be incorporated at lower levels in the mix (more on this in Chapter 24).

3.6 Microphones and Their General Traits

There are many exceptional manufacturers of microphones that are well suited to orchestral and classical music recording. At the risk of upsetting several makers of high-quality products, it can be said that for the main microphone system, the four most commonly used brands in classical music recording are DPA, Neumann, Schoeps and Sennheiser. These companies have been in the business of designing microphones for use in main systems for a very long time. Of course, there are many other brands of microphones that are very high in quality and that are excellent recording tools, such as Audio Technica, AEA, AKG, Beyer Dynamic, Microtech Gefell, Royer Labs, Sanken, Shure and Sony, and the list goes on.

DPA or Danish Pro Audio is a company that grew out of measurement microphone company Brüel & Kjaer (B&K), when operations were split between commercial recording and measurement equipment. The most common omni in their line is the 4006, a small-diaphragm condenser that runs on phantom power and was first released in 1986. The newest designs are transformerless on the output, which provides for higher resolution and a more "open" sound, but slightly reduces the maximum recommended distance between the microphone and preamplifier. Originally designed for taking measurements, DPA omnis are known to be very natural sounding and transparent, with excellent sensitivity and high- and low-frequency extension. They also have a high SPL (sound pressure level) rating, so the capsule will continue to operate normally in a very loud environment. As you can imagine, these microphones will yield excellent results in a beautiful hall, just as they will faithfully reproduce the worst characteristics of a poor acoustic space. In extremely reverberant spaces, the 4006 will need to be placed slightly closer to the source than other brands of omnis. DPA 4041 is a large-diaphragm fixed omnidirectional microphone with both a rising response and more directional polar pattern into the high-frequency range, making it an excellent tool for immersive audio capture.

Neumann has been making microphones in Germany for over 85 years and was acquired in 1991 by Sennheiser. In the early days of stereo recording the famous M50 was used by many recording companies, including Decca, and it is still very popular in orchestra recording for film scores. Updated versions of older microphones also exist, such as the M150 and most recently, the new M49V. Its most popular small condenser "omni" for many years was the KM83, followed by newer versions KM130 and the current model KM183. This is an extremely popular choice of microphone for main systems, especially in Europe. A fully digital version is also available (KM183D).

Schoeps is another German company, which has been providing quality microphones since 1948. Current designs all have interchangeable bodies and capsules, allowing for options in polar patterns and frequency response without having to purchase many complete microphones (DPA is now doing this as well). Their range in dedicated omnis includes the MK2, MK2S and MK2H, all with variations in high-frequency response.

Sennheiser, who began operating in the 1940s as Lab W, have their own line of small condenser microphones. The MKH20 omni and a new, more compact modular design, MKH8020 – with a frequency response of up to 50kHz – are also available in digital versions. The MKH 800 is a dual-capsule design, and the 800 Twin offers separate outputs from two back-to-back capsules, allowing for an adjustable polar pattern during the mixing phase of the recording.

All four of these companies make excellent microphones across a large range of polar patterns, and the choice of which model or brand to use comes down to the subjective preference of the user. Any comparisons should be made with care, and without bias whenever possible. For instance, it has been proven time and again that in comparing two different microphone models positioned at the same distance and height from a source, one of the two microphones will be determined to be superior for that particular placement. A better comparison would be to place each microphone (or microphone system) at an optimum distance and height for the model in question, and then compare the results. An effort should be made to at least match the direct-to-reverberant content of each system, so that a fair and proper comparison can be achieved, resulting in a more accurate evaluation.

3.7 Chapter Summary

- Microphone design varies in both transducer characteristic and polar pattern.
- Engineers must become experts in these differences in order to properly utilize these tools.
- As stated in the previous chapter summary, microphones and loudspeakers are the recording engineer's most important tools.
- Standard stereo microphone techniques must be well understood so that a decent stereo recording of an acoustic source can be made.
- Listen to the piano recording on the companion website to compare various stereo microphone techniques.

https://routledgetextbooks.com/textbooks/9781138854543/audio-examples.php

References

1. Rumsey, F., and McCormick, T. (2014). *Sound and Recording: Applications and Theory*, 7th Ed. Burlington, MA: Focal Press.
2. Eargle, J. (2005). *The Microphone Book*. Burlington, MA: Focal Press.
3. Martin, G. (2011). *Introduction to Sound Recording*. Online only. Website: http://www.tonmeister.ca/main/textbook/index.html.
4. Woram, J. (1982). *The Recording Studio Handbook*. Plainview, NY: ELAR Publishing, p. 12.

Part II

Recording Orchestra

Chapter 4

The Orchestra and Its Various Iterations

This chapter is an overview of how various conductors and orchestra management choose to physically arrange the orchestra members on the stage. For various reasons, some prefer to split the first and second violins left and right, others like them to sit together on one side. Many orchestras sit with cellos on the outside, and others prefer to place the violas in that position. Physical constraints of stage size versus the number of musicians required to perform each piece frequently play a role in stage placement decisions. These seating arrangements each have different effects on the recorded sound, so it is important that they be discussed. Such a large ensemble needs to be placed on the stage so that the balance of each section can be controlled by the conductor as much as possible, and by the microphone setup.

Newcomers to orchestra recording will find that studying the written musical score can be a big help in understanding each section of the ensemble. Even a novice musician who only learned to read treble clef as a student can begin to negotiate a full score, and with some practice it can be a very useful roadmap during a recording session. A recording engineer with orchestral playing experience on their résumé may have some advantage over others, having witnessed the inner workings, although this chapter should serve in catching up those who have missed out on the experience. It should be noted that all references in the text to left and right sides are from the conductor's perspective, which is the same as that of the audience.

4.1 Standard Orchestra Seating

Much has changed in the basic makeup and size of the orchestra since the late 17th century, when composer Jean-Baptiste Lully died from a gangrene infection, after having brought his heavy conducting pole down on his foot by mistake [1]. Certainly Beethoven should receive the credit for standardizing the classical orchestra makeup that includes double winds and brass [2]. From the audience's perspective, it may seem that each orchestra has its preferred stage setup, but in fact it is generally the conductor who ultimately decides how to configure the sections of the orchestra on stage. Even the most well-established orchestras will change their configuration for a newly appointed conductor, or guest conductor, so it cannot be assumed that the sections will always sit in the same place on stage throughout an entire season. The conductor may also decide to change the setup based on repertoire, honoring various composers' intentions in the score (either specific or implied), from Vivaldi to Bartók and Mahler. One common theme in all of the standard seating plans is that the louder instruments such as brass and percussion are positioned at the back of the stage, whereas the strings occupy the front area of the stage closest to the audience, with the woodwinds in the middle. The biggest differences are observed in the seating plans of the string

DOI: 10.4324/9781003319429-6

sections. Several standard string layouts have become common over the years, optimized for different applications.

4.2 Traditional or Standard String Seating

What is now known as a "traditional" orchestra layout is in fact a fairly modern development that is most commonly credited to conductor Leopold Stokowski and his years with the Philadelphia Orchestra, beginning at some point in the mid 1930s [3]. In this iteration the first and second violin sections are located on the conductor's left side, and the violas and cellos are on the right side with the double bass section. Many North American orchestras have adopted this setup, with the cellos sitting on the outside (see Figure 4.1). A slightly less common version of this configuration places the violas on the outside right, and the cellos inside, as in Figure 4.2. In researching this I came across a diagram written in German that refers to these two setups as "American" and "American Alternate".

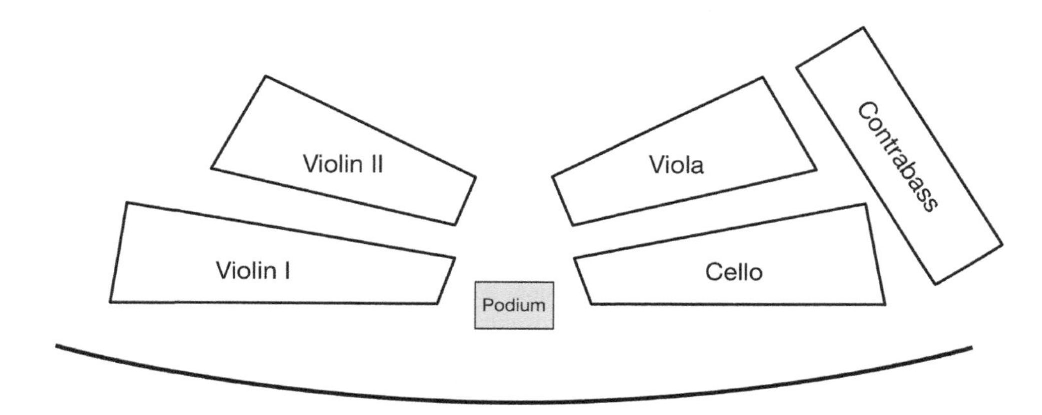

Figure 4.1 Standard String Seating

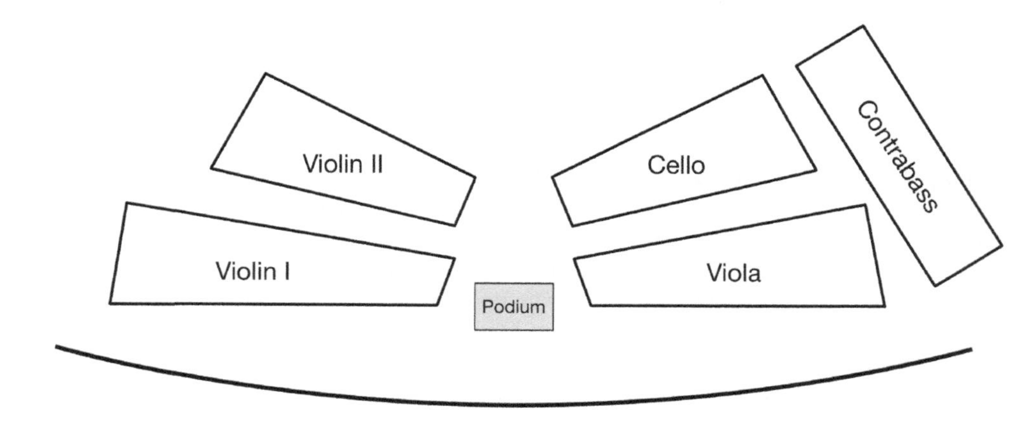

Figure 4.2 Alternate to Standard String Seating

Looking through the musical score of any standard symphonic repertoire will provide a better understanding of orchestral seating plans – for example, it may become clear from the score that cellos and basses frequently "double" the same part, so they tend to be placed near one another.

4.3 Separated or "Split" First and Second Violins

Seating violins to the left and right of the conductor is considered by many to have been the original standard orchestral seating from the late Baroque and early Classical period through to the late Romantic era. This setup remains as the most common configuration for opera orchestras, both in Europe and North America. Very small orchestra pits, however, often place the violins and violas on the left, cellos center and the woodwinds and brass on the right. Many composers have written pieces that take advantage of the left/right spread of the violin sound (Wagner, Mahler). In this layout from left to right, the strings are seated as such: first violin, cello, viola and second violin, with the basses on the left behind the cellos (Figure 4.3). One German source, the Hochschule für Musik Detmold, calls this a "European" setup. Of course, these labels can quickly come undone when a European conductor is hired to lead an American orchestra, or an American conductor who has studied abroad adopts the "German" seating featuring the split violins.

There are many supporting arguments as to which of these setups is best, and it mainly comes down to preference. Certainly, the split violin setup can create an interesting left/right effect when a composition includes musical phrases that are handed back and forth between the violin sections, but there is a trade-off in presence when the second violins play with their backs facing the audience. They will sound somewhat darker than the first violins, and noticeably less powerful in this exposed position. Also, the issue of ensemble must be considered; as the two sections generally play similar parts separated by an octave, it will be more difficult to achieve rhythmic precision between the two sections when they are spread across the entire width of the stage.

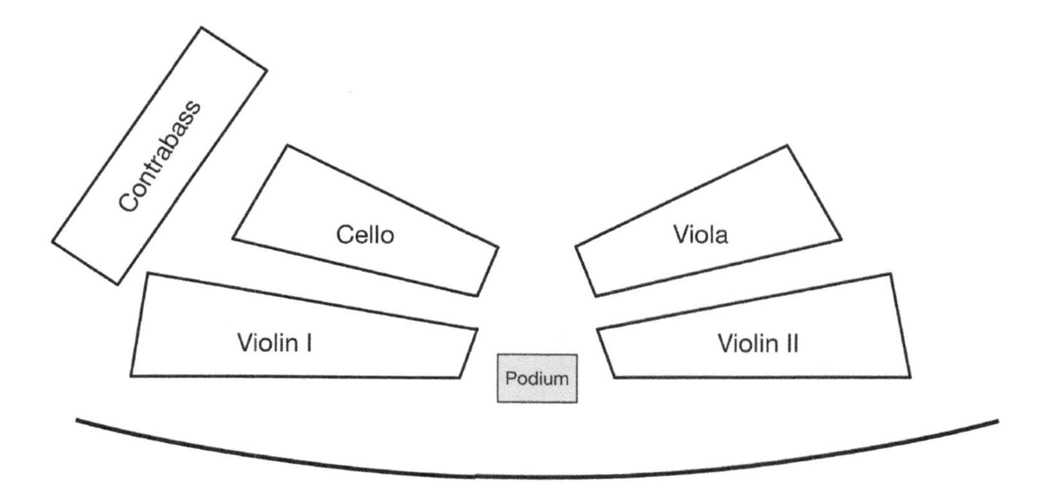

Figure 4.3 Split Violin Seating

Across the general repertoire, there are as many interesting melodic counterpoints to be found between the high and low strings (violins vs. viola or cello) as there are between the two violin sections, so that the effect of interplay between left and right sides of the orchestra is equally inherent within the standard orchestra layout. That being said, it really shouldn't be necessary for two sections sharing or trading melodic lines to be presented from different sides of the recording. As each new seating plan is adopted there will always be some musicians who welcome the change, and others who will complain. When considering a normal size orchestra comprised of 60 to 100 musicians, it quickly becomes clear why the conductor generally makes this decision on their own.

In terms of recording configurations, the standard layout provides for the most even presence across the string sections, as all the instruments face the main pickup. Seating the first and second violins together results in strengthening the power of the upper strings and improves both rhythm and intonation between the two sections. Positioning the cellos on the outside helps to achieve even presence from low to high strings, which is harder to do when the violas sit on the outside, and the cello section is positioned to the right of center, and farther away from the main microphone array. No matter which configuration is in play, the recording engineer must adjust their technique and provide the best possible balance and presentation of the string sections (more on this in Chapters 6 and 7).

As far as audience is concerned, the cellos on the right side radiate quite well into the hall, whereas violas or second violins on the right side will mostly radiate away from the audience, especially in the upper-midrange or presence range. Also, it is important to note that only those audience members sitting near the stage will experience a left/right effect from having split violins, as the direct sound of the orchestra gives way to mostly reflected sound after a distance somewhere around 6–12 rows back into most halls behind the conductor. Any farther back than that, for the average concertgoer the localization of sources within the orchestra is mostly supported through visual cues, and this certainly holds true in the case of the strings.

4.4 Principal Players and Concertmaster

Each string section has a principal or section leader who sits at the front of the section, on the side closest to the audience, so that any solos are easily heard and hopefully seen, and rehearsal dialog with the conductor and other principals is easy to execute. It is also important to note that string players commonly share one music stand between two, and that the "inside player" is always responsible for turning pages (see Figure 4.4).

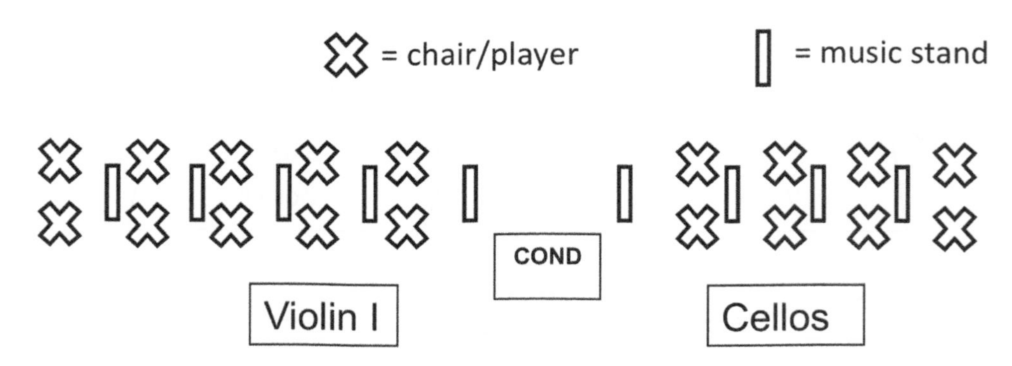

Figure 4.4 Chairs and Stands for 10 Violins and 8 Cellos

The concertmaster is "second in command" after the conductor and is also the principal player of the first violin section, sitting to the conductor's left. The concertmaster also suggests or decides upon bowings for the entire string section. They typically come on stage before the conductor (and soloist) and receive their own applause, before leading the general tuning of the orchestra. As guest conductors come and go, the concertmaster usually has the best overall impression of the orchestra and its personnel and can be very helpful in discussing many issues such as basic strengths and weaknesses as well as the overall sound of the ensemble.

4.5 Woodwind Section

The woodwind section is placed in the center of the ensemble, directly behind the strings (see Figure 4.5). As far as the double woodwind convention goes, the seating is standard – flutes and oboes always in the front row, with clarinets and bassoons behind. Principals sit on the insides, so that Flute 1 is next to Oboe 1, and the same applies to the second row. Additional winds as needed sit on the outsides of each section (English horn, bass clarinet, etc.) as in Figure 4.6. Saxophone, when needed, is quite often seated to the left of the clarinet section (or in the bass clarinet seat when only two clarinets are required). Risers are often used to help the woodwind sound project through the string sections into the hall, with the second row slightly higher than the first row, or only the second row on risers at all. In terms of recording, risers can help position the woodwinds slightly closer and with a clearer path to the main microphones, although having woodwinds on the floor in a good hall can also work well (see Chapter 8 for more information).

There is no overall section leader of the woodwinds; rather the four principals make up a sort of committee, which seems positively diplomatic at first look, although differences of opinion arise on issues such pitch and phrasing, which may be hard to resolve.

4.6 Brass and Percussion

The brass section technically includes the trumpets, trombones, tuba and French horns, although the horns are frequently treated as a separate section. Typically, the principal trombone leads

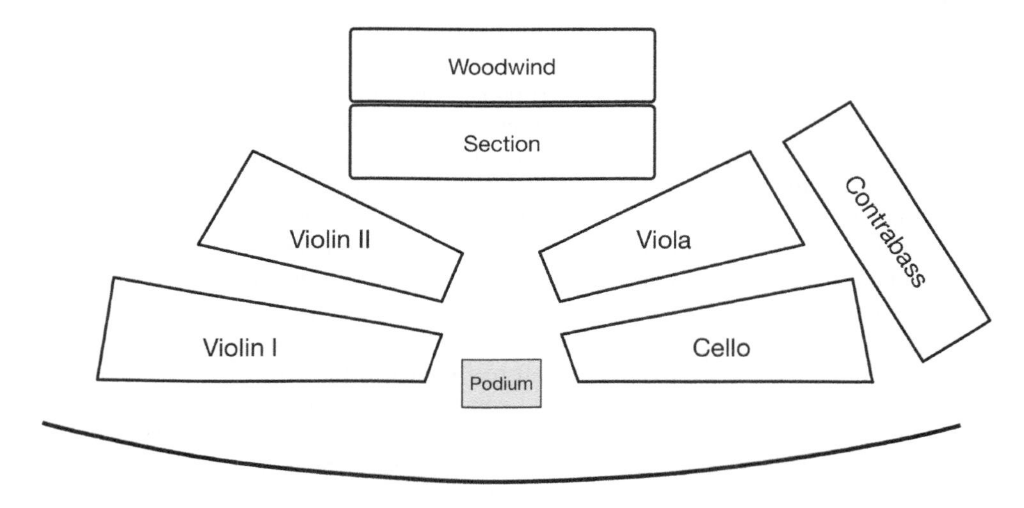

Figure 4.5 Woodwind Section Stage Position in Relation to the Strings

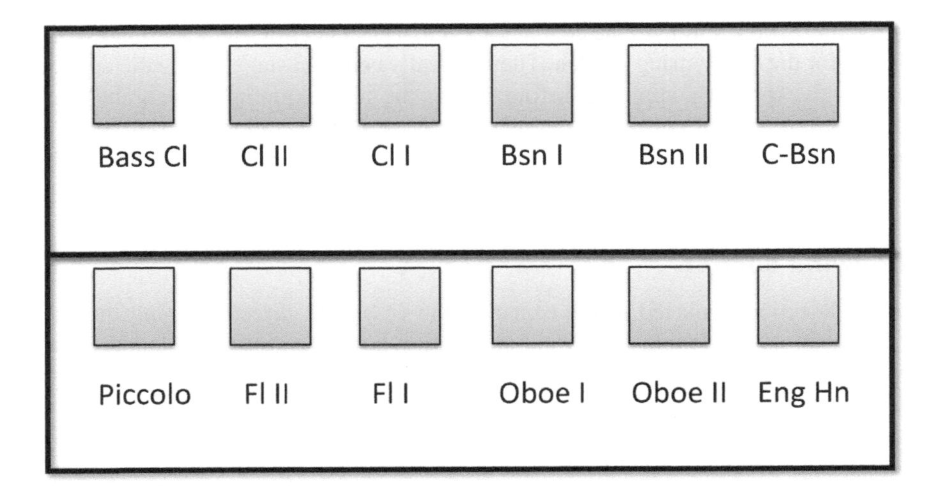

Figure 4.6 Woodwind Seating Showing Principals in Center Chairs

the low brass, and principal trumpet is the leader of the entire brass section. Seating plans vary, although it is common for the principal trumpet and trombone to be sitting together in the middle of the brass section. The easiest way to find out where a player is sitting is to look at the title on the cover of the folder on each music stand, or by simply asking the librarian or stage manager. This is important when placing support microphones within the sections (see Chapter 8). Stage layouts for brass and horns depend greatly on available space and number of players – on a wide enough stage the brass can be seated in one line, starting with trumpets in the center or slightly right of center, followed by trombones and tubas spreading to the right, with the horns on the left side or in the center in a third row behind the woodwinds, based on the conductor's preference (Figure 4.7). A narrower stage might require that the trumpets and trombones sit in two rows on the right-hand side, with the horns either on the left side or centered behind the woodwinds (Figure 4.8).

Looking at the score, it can be seen that horns are normally paired up in a specific manner. While they sit left to right as IV, III, II and I, it is Horn I and III that mostly play in the higher registers, where Horn II and IV tend to be assigned lower parts. This is because two horns were standard in early Beethoven symphonies, and as the forces grew in later repertoire, two more horns were added to double or otherwise strengthen the original pair. Composers in later works frequently call for eight horns or more, for example Richard Strauss and Gustav Mahler. Orchestras that frequently perform larger-scale works will have an assistant principal for both the trumpet and horn sections, so that two players might split up the task of covering very demanding high parts in a tutti section or important solo passages. In the trumpet section this role is quite often covered by the third trumpet, as Trumpet II tends to make a career of blending with the first trumpet in a lower register. The assistant horn is normally an additional member of the section, sitting next to the principal.

The percussion section generally fills the left rear corner of the stage, with the timpani positioned closer to the center, where there is good communication with the trumpets (Figure 4.9). Having all of the percussion together is preferred by the players but is less desirable for recording – more on this in Chapter 8. If and when the horns are placed behind the woodwind section, the timpani may remain centered behind the horns, or they may be placed in front of the percussion to

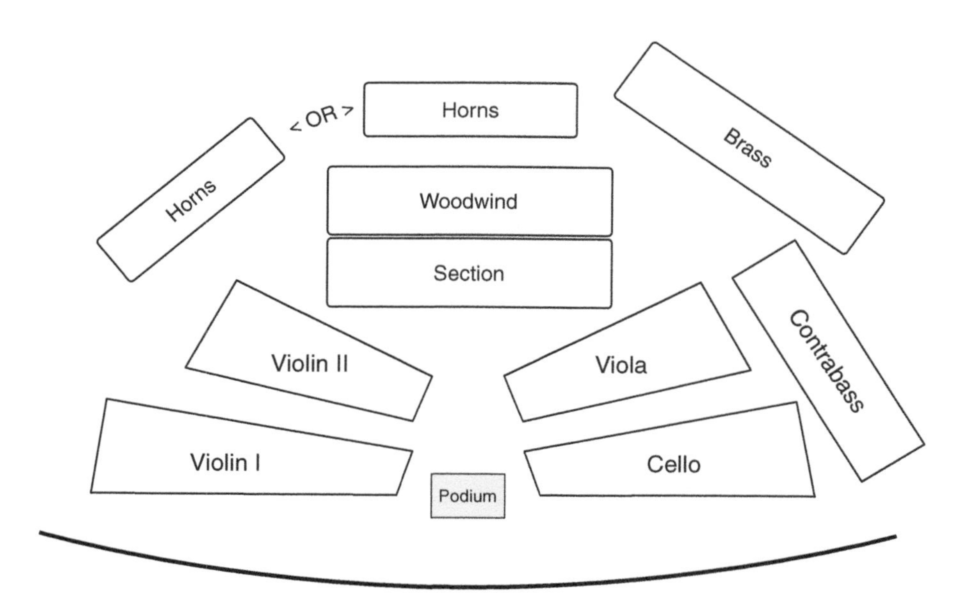

Figure 4.7 Example of Horns and Brass on a Wide Stage

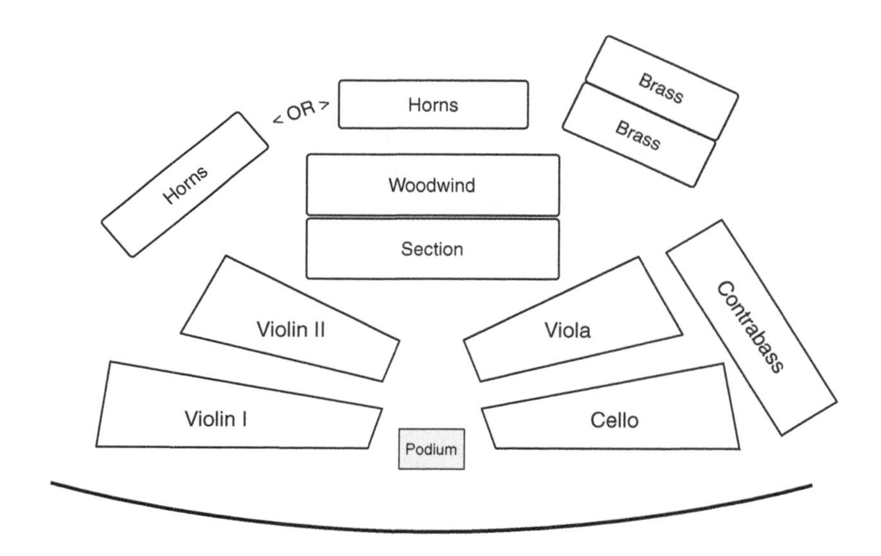

Figure 4.8 Example of Horns and Brass on a Narrow Stage

the left of the woodwinds. As more space is needed, the percussion may also take up space on the right side behind the brass. In professional orchestras, playing the timpani is considered a specialty role and as such the "timpanist" rarely plays any other percussion instruments. The principal percussionist is responsible for assigning the parts to each section member.

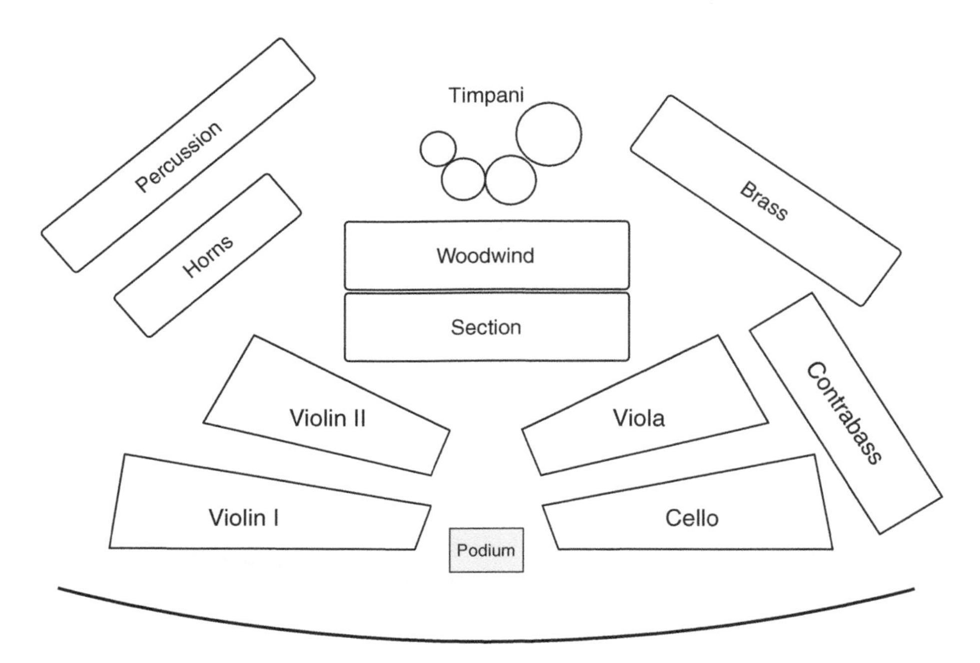

Figure 4.9 Typical Positions for Percussion and Timpani

4.7 Harp, Celeste and Orchestral Piano

There is no standard position for these instruments within the orchestra. A common solution is to place them "wherever they fit", and to hope for the best. Typically, there are more large instruments on the right-hand side of the orchestra, so it tends to be easier to fit these instruments on the left side, normally at the back of the violins (Figure 4.10). Orchestral piano and celeste are often considered part of the percussion section, so the proximity to that group can be helpful for communication. The harp is quite a narrow instrument, and at times it can be successfully placed in between string sections (normally on the left) and about halfway toward the front so it has a better chance of being heard. Orchestras that seat the horns in the center behind the woodwinds usually have room for harp or piano to the left of the woodwind section. Piano and celeste parts are quite often written so that one person plays both instruments, requiring them to be placed next to each other.

There are several other placement options for these instruments during recording sessions, which will be covered in detail in Chapter 9.

4.8 Planning the Stage Layout

The entire stage setup needs to be addressed well before the recording session begins, as even small changes to the stage layout can take a great deal of time and negotiation with the musicians. This is normally achieved with the help of the stage manager, who is more than familiar with the personalities of each orchestra member. Ideally the orchestra rehearses in the same configuration that will be used for the concert or the recording session, so that everyone has time to adjust to any alterations in the seating plan. Major seating changes between pieces must be planned with

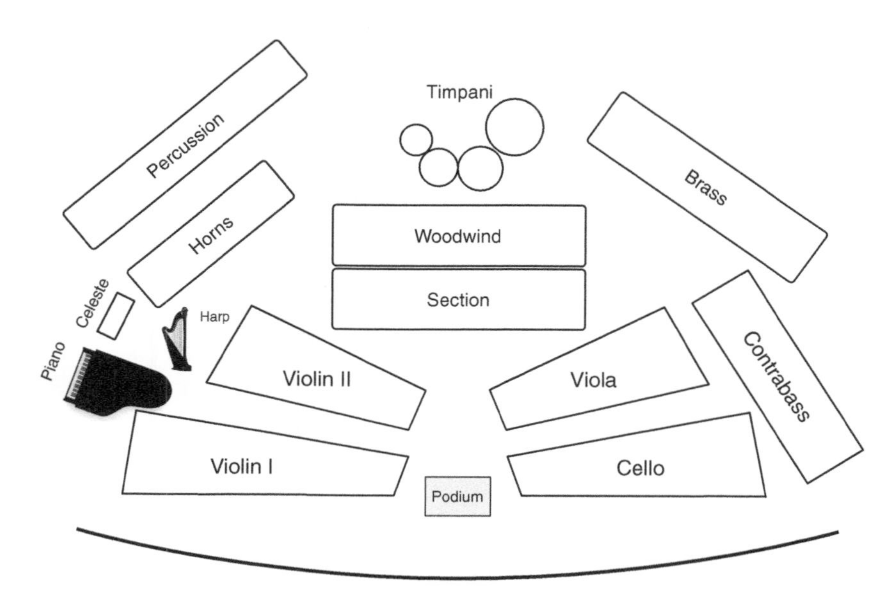

Figure 4.10 Example Placement of Harp, Celeste and Piano Within the Orchestra

efficiency, so that time isn't wasted moving instruments and chairs during the recording or performance. Questions will need to be raised, such as "can the celeste remain on stage during the Overture, even if it won't be needed until the second piece?"

A great place to start investigating how a certain piece might be best laid out on stage is to look inside the score. On the left side, across from the first page of music, there is usually a listing of the instrumentation, including how many of each instrument is required and every percussion instrument used, and the convention is to list each section in score order from top to bottom. You don't need to be an expert at score reading to use this handy chart. Figure 4.11 shows an example of a typical instrumentation list, from a fictitious score.

It is quite rare in the score for the string forces to be specified in exact numbers; rather, it is up to the conductor (or artist administrator) to be familiar enough with the piece to know how many string players are needed for each section to balance out the forces. For example, a Mozart symphony uses less than half the brass and woodwinds of a large work by Tchaikovsky, so the size of the string section will be adjusted accordingly. Also, some halls may work better with more or fewer low strings, so that the numbers remain variable as needed.

String sections (and in fact all sections of the orchestra) can be easily quantified by a series of numbers, representing a player count for each section in "score order" – first violin, second violin, viola, cello and double bass. Therefore, a typical early Classical string section might be 10-8-6-4-2, whereas a Mahler symphony might require 16-14-12-10-8 to balance the massive brass forces and triple woodwinds. The numbers are usually even, since pairs of players share a stand, but occasionally there may be an odd number somewhere, for example, 10-8-5-4-2: in this case, one viola would have their own stand and would turn their own page. This instrumentation "shorthand" can be a quick way to track the total numbers for each piece on a program. The chart in Figure 4.11, for instance, can be reduced to a few lines that document everything needed for stage preparation. With ease, one can quickly see how many chairs and stands are needed for each section, beginning with woodwinds, followed by brass, and finishing with a breakdown of the string section:

2 Flutes	3 Trumpets
Piccolo	2 Trombones
2 Oboes	Bass Trombone
English Horn	Tuba
2 Clarinets	Timpani
Bass Clarinet	Percussion*
2 Bassoons	2 Harps
Contra Bassoon	Celeste
4 Horns	Piano
Strings	
*Percussion: Bass Drum, Snare Drum, Cymbals, Triangle, Tambourine, Xylophone	

Figure 4.11 Example of the Inside Page of a Musical Score Showing Instrumentation

3-3-3-3 (Fl Ob Clar Bsn)
4-3-3-1 (Hn Tpt Tbn Tba)
T + 3 P (Timp + three percussionists)
2Hp Cel Pno (Two Harps, Celeste and Piano)
14-12-10-8-6 (I II Vla Vcl Cb)

4.9 Chapter Summary

- The orchestra itself is a complex and multifaceted sound source.
- Seating configurations will change based on preference, and space limitations.
- The score can be very informative as to what is happening onstage.
- Advance planning is needed to properly prepare an orchestra recording!

References

1. La Gorce, Jérôme de. (2002). *Jean-Baptiste Lully*. Paris: Fayard.
2. Kern Holoman, D. (2001). "Instrumentation and Orchestration: 4. 19th Century." *New Grove Dictionary of Music and Musicians*, 2nd ed. New York: Grove.
3. Opperby, P. (1982). *Leopold Stokowski*. Tunbridge Wells: Midas Books.

Chapter 5

The Hall as a Recording Venue

The concert hall or recording studio should be thought of as one of the instruments in classical music recording, as the most common techniques capture the sound of the instruments and the room as one element. It is important to assess the sound in the hall before listening to the resulting sound coming over the microphones, so that the engineer has a mental snapshot of the acoustic "signature" of the orchestra in the hall. The goal of the recording is to replicate the sound of the musicians in the room, with the intention of improving any shortcomings of the space through the careful placement and balancing of microphones. In extreme cases, the hall itself might be physically altered to improve its acoustic properties, by employing a stage extension, placing sheets of plywood over the audience seats or hanging theater curtains off a balcony to reduce excessive amounts of reverberation. Large spaces, other than halls and studios, may also be considered as suitable recording venues. Many churches have excellent acoustic characteristics and are generally acceptable for recording, as long as there is adequate isolation from the outside surroundings.

It is important to keep in mind that the artist is expecting the general presentation of the recorded sound to match their experience on stage. The most flattering comment for the recording team to hear is that the artist prefers the sound in the control room to that on stage, although it is more common to hear that it sounds "exactly the same", even though that would be quite impossible with stereo reproduction. Great care should be taken, however, to avoid hearing the opposite opinion – that the artist prefers the "live" sound on stage and is having trouble transitioning from what they hear on stage to the sound in the playback room. In this case, a change in the microphone placement or a significant correction of the balance is in order. Sometimes this problem can be partly attributed to the control room setup, as discussed in Chapter 14. Here, the advantage of headphone listening can be brought into play when an artist needs to assess the recorded sound in isolation from the control room acoustics. For more information on the acoustics of concert halls, these publications go into great depth [1–4].

It is not very often that a recording engineer will have much to do with choosing the venue for an orchestral recording. The orchestra's schedule, the recording budget and practical issues, such as the location of the orchestra, and the ability to transport instruments and equipment, usually dictate the decision. Once in a while, the recording team may be asked for their opinion regarding one hall over another, but it is uncommon for the engineer to be asked, "What is your preferred hall for this project and the proposed repertoire?" The best advice here is that an engineer should develop techniques and strategies to deal with less-than-ideal recording spaces, since these will be the most common venues available.

Many orchestras are obliged to record in their own hall, even if there is a better hall nearby. Quite often the orchestra is the main tenant of the hall in which it performs and is already renting the space or can use the space for recording sessions at a discounted rate. In fact, it is more

DOI: 10.4324/9781003319429-7

common for orchestras to record their live concerts than to engage the musicians in very expensive recording sessions, so the recording location and time must follow both the concert schedule and the performance venue. This can be very limiting, so it is the burden of the engineer to remain adaptable and to make the best of the circumstances. The next section is an overview of the various types of orchestral recording venues one might encounter when embarking on location recording.

5.1 The Concert Hall

Many great concert halls exist in the world today, some of which were built hundreds of years ago, while others were completed in recent months. The earlier halls were designed mostly under aesthetic criteria, and those halls that ended up with exceptional acoustics were probably a result of chance and the use of excellent materials. Ornate use of wood and plaster materials in the designs provided for substantial diffusion, which influences the quality of the reverberation and helps to breakup distinct reflections or "slap" echoes. When surveying a hall for a potential recording, which should be done in advance of any recording (and whenever possible), there are several important criteria to consider:

Clarity of sound on stage: Many halls will have a moveable shell that can be adjusted, to help project the sound of the back of the orchestra forward into the hall. This aids in balancing the back half of the group with the strings in the front, which are closer to the main microphone system. Ceiling clouds help to reflect the stage sound into the house and, along with the stage walls or shell, they help the musicians hear each other more clearly, improving rhythmic ensemble and leading to a higher quality performance in general.

Stage extension and orchestra pit: Many halls have a "pit" section where the orchestra can be placed below the stage level, leaving the main area free for theatrical scenery and other performers, such as opera singers or ballet dancers (see Figure 5.1). The pit floor may be moveable via motors, or the pit opening may be covered with removable stage flooring when not in use.

For symphonic concerts, the pit is normally brought up to the level of the floor in the house, and several rows of chairs are added to accommodate more audience seating. As required, an adjustable pit can also be raised up to the stage level, offering another 2.5–5 meters (8–16 ft.) of depth, which can help extend the ensemble out into the hall (as shown in Figure 5.2). This can be advantageous in live concerts when a piece requires vast forces, such as a late Mahler symphony with large orchestra and chorus. A stage extension can allow the entire string section to be moved out into the hall. This may be useful in a space where the proscenium is actually inhibiting the projection of the string sound into the hall and the microphones, or at least causing a difference in acoustic character between those string players sitting outside of it, and those sitting farther upstage toward the rear wall of the stage (and therefore caught under the proscenium). A stage extension can be an advantage during a recording session, even if it is only employed to give more wood surface in front of the strings, which can improve the sense of power in the sound by adding additional early reflections into the hall and the main microphones. It is also helpful to have easy access to the front of the orchestra at the same height as the stage, so that quick adjustments to microphones can be made without having to negotiate a path through the musicians to reach the area around the podium.

Proscenium: this is the opening where the stage meets the "house" or audience area, and where the curtain drops down to close off the stage area from the rest of the hall. This is typical of theaters used for opera and ballet so that scene changes can be achieved without the audience witnessing the transition. Many halls built for orchestral performance do not have a proscenium. When one exists, however, it should be evaluated as to how it affects the sound on stage, and how the sound projects differently into the hall depending on whether the sound source is in front or behind this divide.

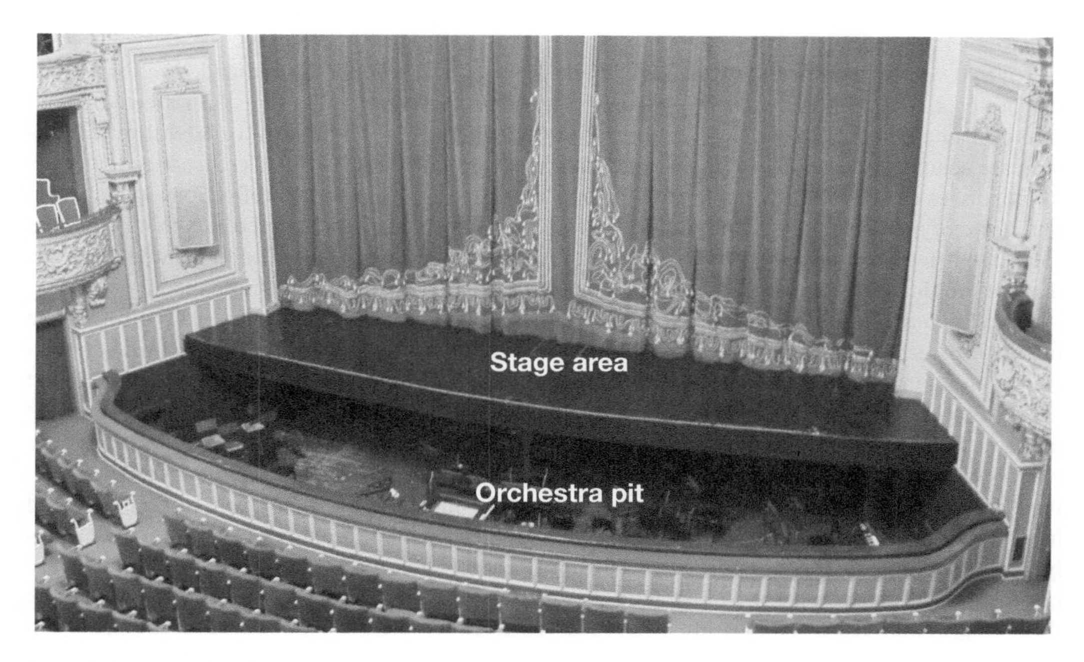

Figure 5.1 Image of an Orchestra Pit

Figure 5.2 Image of a Stage Showing Proscenium Arch and Stage Extension

Credit: Ben Huser, used with permission

Orchestras typically spend a great deal of time trying to position the musicians on stage in relation to the proscenium so that the best possible sound and balance is achieved.

Ceiling height: Halls with low ceilings can be problematic, both for the audience and for recording projects. The best halls are designed with ceilings high enough so that the reflected sound arrives at the listener from the sidewalls before the ceiling. These sidewall or lateral reflections increase spaciousness and help to generate listener envelopment. This effect is reproduced over the microphones as well. A low ceiling will increase the strength of vertical reflections and shorten their time of arrival. When ceiling reflections are stronger (or earlier) than those coming from the sides, the listening experience can be rather unpleasant. In the case of a high ceiling, late reflections may be problematic unless an appropriate amount of ceiling absorption is employed.

Quality and amount of reverberation in the hall: Very dry halls can make it hard to get the instruments or sections of the orchestra to blend. Also, a lack of "bloom" in the decay of the hall can be anticlimactic during the louder "tutti" sections of a piece, or pauses in the music where the hall has a chance to be heard on its own. Long lines or solos in slow passages may not "float" as much as they should. Reverb acts as a cushion, and without it a performance can fall flat. Very reverberant spaces, however, can also be problematic. It may be difficult to get enough clarity in the direct sound, and if the quality of the reverb is highly unpleasant it will be very difficult to avoid having it in the recording. In this case, simply placing the main microphones closer to the ensemble will not always yield the appropriate balance of direct and reverberant sound – imagine a very close-sounding recording with a distracting "bathroom", or other uneven room sound, looming in the background. Newer designs might include some retractable banners or screens along certain walls of the hall, which can help shape the length of the reverberation. A helpful general guideline here would be to err on the live side during a soundcheck/dress rehearsal and adjust from there. It is also good to at least consider the advice of those who use the hall on a regular basis when deciding on a starting position for these moveable acoustic treatments.

Raked or flat floor in the house: Raked floors gradually rise in elevation as they extend toward the back of the hall. This design is quite common, as it provides for better sight lines and more direct stage sound for the audience, to the back of the house. The floor design has less influence on the recorded sound of a hall, except that a raked floor will require fewer absorptive materials on the walls and ceiling.

Balcony: Having a balcony included in a hall can influence the length and complexity of the decay of sound in the room, or the reverberation "tail". As sound is reflected under a balcony, it tends to be "returned" to the main room well after the reflections from the main area of the room. The effect can be a longer decay time, or even a subtle "second wave" of decay, with a darker timbre, and can be a very interesting attribute of a hall. On the other hand, a very late return of sound from under a large, deep balcony might be confusing to the listener, and may clutter an otherwise "clean" hall sound in a recording.

Background noise/isolation: Of course, a quiet space is a priority for recording classical music, due to the potentially large dynamic contrast from extremely soft to very loud passages. Background noise can therefore be a major issue. Quite often a good-sounding hall will have a very noisy set of power transformers or lighting dimmers backstage, which might bleed onto the stage. Onstage lighting should be auditioned in case it produces hum, buzz or a high-pitched whine. Air-handling systems for cooling and heating might also be too noisy to be operating during a recording session, and it should be determined whether or not the system can be controlled during recording. Very often there are minor problems that can be solved easily – a refrigerator in an offstage room that might need to be unplugged, for instance. Traffic noise can be a steady-state issue depending on the time of day, whereas less frequent sources of noise might be accommodated, such as ambulance sirens near a hospital, a passing truck or children exiting a schoolyard in the late afternoon. Some of these artifacts can be successfully filtered out, but it is preferable to eliminate them at the source.

Location of an isolated listening room: Although this is a low priority when it comes to choosing a hall, it is always good to inquire ahead of time about a room for listening that is quiet and well isolated from the stage. Even headphone listening requires a separate room from the hall in order to properly judge microphone positions and balance. A noisy listening room will also make it difficult to discern between background noise on stage and any rumble or hum that exists only in the control room. Some reorganization of musicians and staff may be required if a certain dressing room or office space is needed, so it is best to agree on the location of the control room well before the recording session.

5.2 Hall Designs

The basic design and shapes of a hall can have a great influence on its acoustic footprint. It is important to take note of this during a survey – here are a few common designs.

Shoebox shape: some of the most famous orchestra halls in the world are this shape, typically with a shallow balcony or balconies (the Royal Concertgebouw in Amsterdam, as shown in Figure 5.3, Musikverein in Vienna, Symphony Hall in Boston). These rectangular structures

Figure 5.3 The Main Hall of the Royal Concertgebouw in Amsterdam

Credit: Photo: Lauran Jurrius, used with permission

combined with ornate plaster walls provide for a smooth, dense reverberation "tail". Very reverberant halls might require that a large theater curtain is hung in the middle of the house to bring down the decay time and to increase clarity on stage, when they are used for a recording session. This is of course not possible during a live concert recording, although a large audience may help to absorb some of the reverberation.

Fan shape: this design grew out of a desire to increase the seating area close to the stage by widening the audience area and offering better sight lines to the stage. While this helps for speech intelligibility, the fan shape design is generally not suited for music performance, as there is normally a reduced sense of envelopment as compared to a rectangular design. Energy from the stage is quickly dispersed and attenuated into the widening volume of the hall so that it loses its intensity very quickly.

Multipurpose venue: a general rule of thumb is that any venue that is designed to work well for any type of event is usually not very good for any one particular use. It might be too dry for live acoustic performance with orchestra or solo piano, and at the same time too reverberant for amplified pop concerts and musical theater, or too diffuse for lectures or ceremonies including live speech. The concept stems from the budget-conscious desire to build one hall that can be used for many types of events, rather than a complex made up of two or three halls that are each optimized for different purposes. A room that is considered too "dead" for acoustic performance can be improved by placing sheets of plywood over the seats closest to the stage, which increases the number of early reflections and adds power to the sound. This can be an expensive and labor-intensive undertaking. Some theaters have rubber floors for dancers, which will reduce the level and change the timbre of the early reflections – especially in the string sound. It might be worth asking if the rubber floor can be easily rolled up and removed, or if the hall has a wood covering that can be placed over the rubber surface in order to improve the sound.

5.3 The Recording Studio

Large volume recording studios can approximate the sound of a concert hall, except that there is little distinction between stage and hall sound. Two excellent examples of studios suitable for classical music recording are Abbey Road Studio One in London (see Figure 5.4) and the scoring stage at Skywalker Sound in California. The very positive attributes include a completely isolated and climate-controlled environment, state-of-the-art equipment and a large assortment of microphones in-house, with a permanently installed and acoustically treated control room.

Many large studio spaces have variable acoustics, which can allow for an extended reverb time that might help approximate a concert-hall sound. As with halls that have variable acoustics, it is normally best to start with the most "live" setting, and then slowly reduce the reflective surfaces to achieve a more controlled sound as required.

5.4 Churches, Temples and Gymnasiums

It can be hit and miss within this range of makeshift recording venues, and a great deal of work may be needed in order to end up with exceptional results. Some churches are acoustically superior to many concert halls for recording, whereas others can be sonically disastrous due to low-frequency buildup, uneven reverberation and strange echoes (slap, flutter, etc.). Gymnasiums tend to have parallel walls and very little to offer in the way of absorbent materials, and generally should be avoided if at all possible. These venues rarely have adequate isolation from external noise, so that quite often recordings have to be scheduled in the evening or late at night, when the surrounding

Figure 5.4 Abbey Road Studio One, London

Credit: Abbey Road Studios, used with permission

neighborhood is quiet. Churches rarely have decent-sized isolated rooms in close proximity to the "stage" that might be used for listening, and quite often what is considered to be the stage area might be on the small side, or on different levels (transept steps, etc.) In this case a stage platform or a series of risers might be needed in order to position the ensemble on one level.

5.5 Chapter Summary

- Think of the venue as one of the instruments and be prepared to include it as part of the overall pickup.
- Be aware of the change in the influence of the room on the overall sound each time the main microphone position is adjusted.
- Listen carefully to advice from those who use the hall on a regular basis.
- Plan to survey the hall or venue at least a few days before a recording in order to properly prepare for the project.

References

1. Ando, Y. (1985). *Concert Hall Acoustics*. New York: Springer-Verlag.
2. Beranek, L.L. (2004). *Concert Halls and Opera Houses*, 2nd ed. New York: Springer-Verlag.
3. Everest, F.A., and Pohlmann, K.C. (2009). *Master Handbook of Acoustics*, 5th ed. New York: McGraw-Hill.
4. Marshall, A., and Barron, M. (2001). "Spatial Responsiveness in Concert Halls and the Origins of Spatial Impression." *Applied Acoustics*, 62 (2) February, 91–108.

Chapter 6

The Main Microphones

With this chapter, the real "How To" portion of the book begins. Aside from Chapter 2 on how to listen, this is probably the most important section of the publication. In the previous chapter, emphasis was placed on describing concert halls and their various iterations. Stories of the legendary "best seat in the house" are all too common, as every hall has that one magic place to sit where everything comes perfectly into focus and the reverberation is at the optimum level. This "best seat" is generally a matter of debate, as there are invariably five or six of them in different locations around the house, depending on who is being asked. There is also much discussion about how the sound changes from the floor or orchestra level to the first balcony, or how row T on the right-hand side might be described as a veritable "sonic black hole". This is mostly folklore for the patrons to obsess over at the bar during intermission, as none of it really matters for recording. As far as this chapter's material is concerned, the most important position in the room is the location of the main microphone system.

6.1 Main Microphone Systems

Several techniques have been used over the years, with varying success. The following section is an overview of established techniques still in use today. Some overlap will occur here with the information in Chapter 3 (Microphones). Readers are encouraged to try the various techniques and decide for themselves which ones they prefer.

ORTF and *NOS*: It is hard to say if these systems are still commonly in use by the French and Dutch radio engineers, the original developers and users of the techniques. Although these techniques are being used by a fair number of orchestras for archival recording and radio broadcast, it is commonly agreed that these near-coincident techniques are not so popular for commercial recording of classical music – especially where large ensembles are concerned. When used for orchestral recording, these cardioid-based systems are commonly supported with a widely spaced pair of omni microphones in order to achieve better low-frequency response and coverage of a large ensemble (see Figure 6.1).

AB pair: This system is probably the most common technique for commercial classical music recording. If a commercial-quality orchestral recording were to be made with only two microphones (and it can be done), it would almost certainly have to be an AB recording to be really successful in representing all the necessary criteria. A properly placed AB can provide great impact, width, depth and low-frequency response, and with the right microphones, incredible clarity and realism. Spaced omnidirectional microphones as a recording system suffer from reduced mono compatibility, but that is one of the reasons the resulting capture will sound much "bigger" (greater

DOI: 10.4324/9781003319429-8

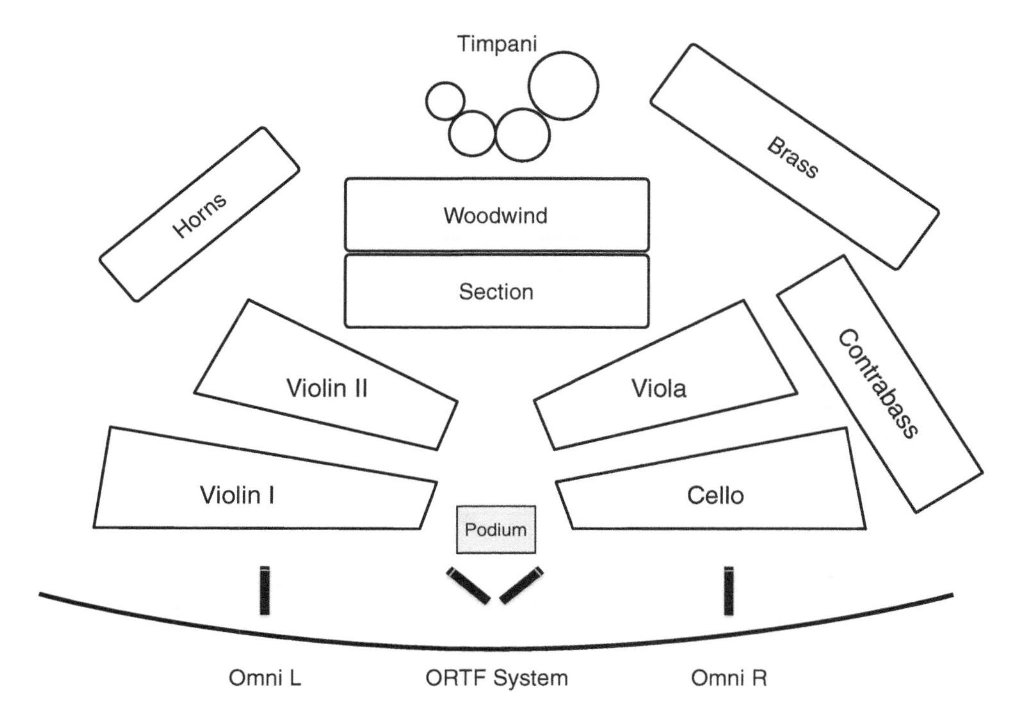

Figure 6.1 Example of ORTF Together with an Omnidirectional Pair

spatial impression) than a system with greater left/right phase correlation. This is due to the differences in time of arrival of signals across the two channels. Single AB has less adjustability in post-production, but when it is placed well it shouldn't need much manipulation in the mixing stage.

It is recommended that newcomers to orchestral recording start with the very simple yet versatile AB technique and get comfortable with understanding how it works before trying to use either a combination of pairs or a Decca tree with an adjustable center component.

Decca tree: This three-microphone array was made famous by English engineers working for the Decca record company in the 1950s, Arthur Haddy and Kenneth Wilkinson [1]. After trying several types of microphones, Decca eventually settled on the use of the Neumann M50 microphone, which was a fixed-pattern omnidirectional design with a tube power supply, except that somewhere along the way they removed the tube electronics and replaced it with transistor or "solid state" circuitry running on phantom power, thereby eliminating the unreliable and bulky power supplies. The M50 utilizes a small diaphragm mounted in an acrylic sphere. The sphere basically creates a polar pattern that is omnidirectional in lower frequencies, and then becomes more directional starting at 1 kHz until it is more or less fully cardioid above 4 kHz. A Decca tree can be built without M50s or the reissue M150s by using a set of omnidirectional microphones fitted with diffraction attachments directly behind the capsule. Examples are the Neumann KM 130 + SBK

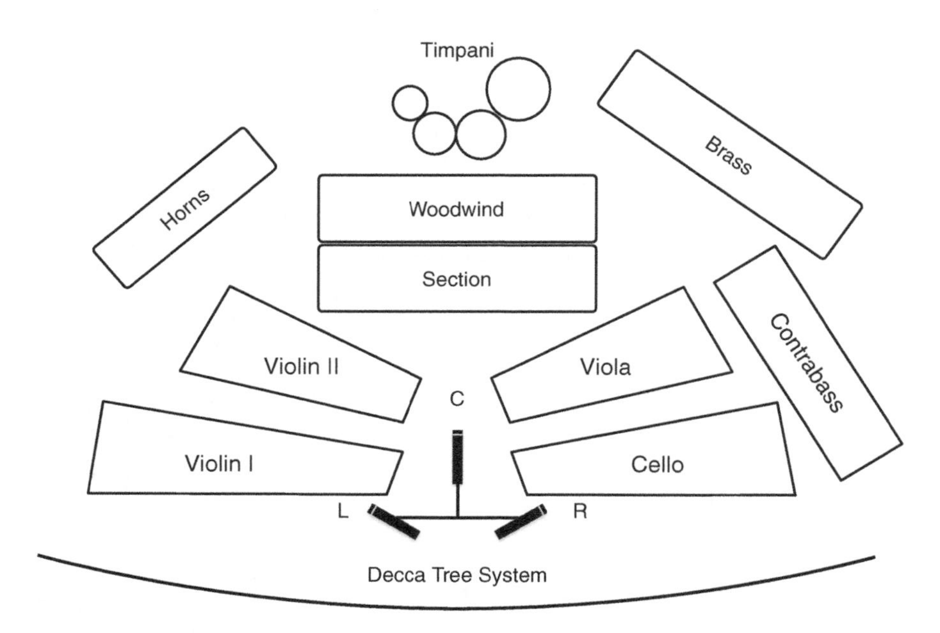

Figure 6.2 Diagram of Typical Decca Tree Configuration

130A, or Schoeps MK2H + KA 40 or the DPA 4006 + APE 40. Sennheiser makes a ring that can be positioned just behind the diaphragm of an MKH20 to create the same effect.

The Decca tree is basically an AB system with an added center microphone, however the left and right microphones are angled outward to achieve better "reach" or presence from the more widely placed sources, because of the high-frequency directionality inherent in the M50 design (see Figure 6.2). The center microphone is used to fill in the middle of the pickup and stabilize the center image, and is placed slightly forward of the L/R pair so that the three microphones appear as a triangle.

The "tree" refers to the very large microphone stand that supports the microphones on three separate "branches" or arms of the stand and can be positioned directly over the conductor's head because of its solid base and long reach. The resulting presentation includes the power and depth of an AB system along with mono compatibility and image stability approaching that of the coincident techniques described in Chapter 3.

Mono compatibility was a major concern throughout the decades of the 20th century because of mono LP releases, mono radio broadcasts and even FM stereo broadcasts; in the latter case, image width can be exaggerated, as center information is reduced because of the manner in which FM transmission is achieved. The Decca tree is still commonly used for orchestra recording and is hugely popular in film score recording because the center microphone can be applied directly to the center playback channel, providing a discrete Left/Center/Right pickup as the foundation of the front channel makeup in surround-sound presentations.

Small ab/big AB: This is a common technique used quite widely in Europe and beyond, and in one configuration a small ab is placed just above the conductor – omnidirectional microphones with a fairly narrow spacing of around 30cm (12") and normally an angle of incidence around 90°. Another technique is to use two omnis in parallel, spaced slightly wider, around 60–80 cm apart

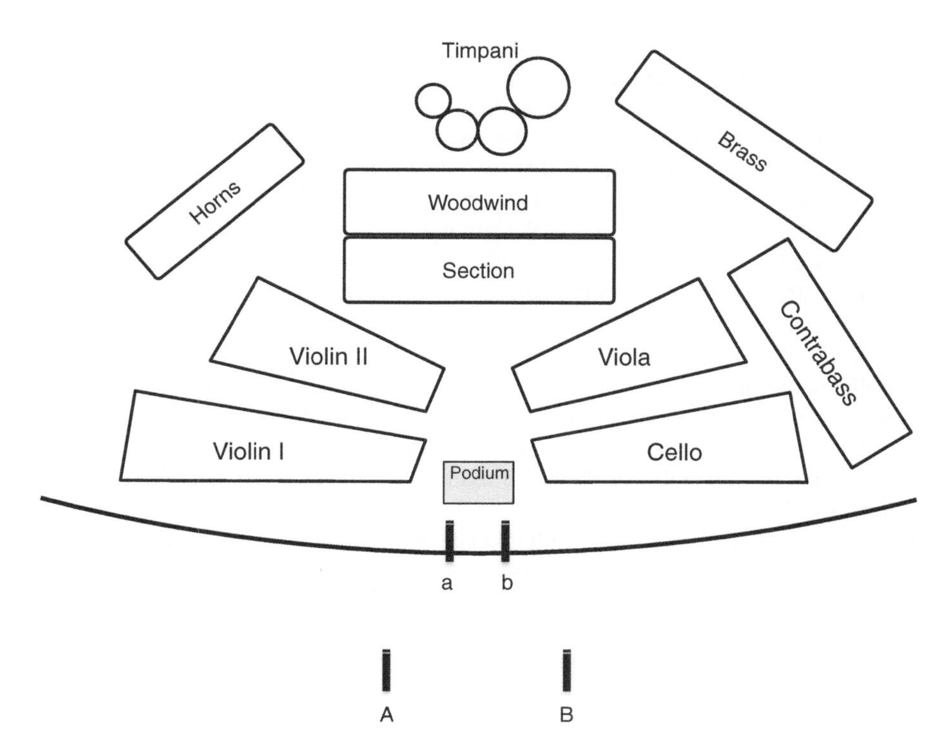

Figure 6.3 Diagram of Two AB Systems Used in Tandem

(24–32"). This is then combined with a big AB, two more omni microphones placed a few meters back and spaced 2–3 m (7–10 ft.) apart, as shown in Figure 6.3.

The small ab provides clarity, impact and detail, whereas the big AB has more blend and hall sound (reverberation). The two systems are then combined at the mixing console, and the balance can yield a more direct or more diffuse result, depending on the levels of each of the two pairs in the mix. Many famous and excellent recordings have been made using this technique, and some recording teams make use of three systems (close, medium, far) in order to further increase options in postproduction. The advantage is that the overall perspective can be easily adjusted after the fact, at the mixing console. The disadvantage, as compared to just one AB system, is a reduction of purity or clarity, as two systems of different distances to the source are combined with an overall phase difference.

Consider a wide AB placed 3 m (10 ft.) behind the closer and narrower ab pair, and the two are combined equally in level. Because of the delay of around 10 ms for sound reaching the more distant AB, the combination of the two systems will result in a slightly veiled sound. This reduction of clarity might actually be beneficial, as it may compensate for the tight and possibly aggressive sound of the small ab on its own. As long as the spacing between the two pairs is great enough there will be very little discernible comb filtering (see Figure 6.5). Imagine a crisp snare-drum solo from the rear left corner of the stage presented over two main pairs, offset by 10 ms and with an image shift due to the spacing of each pair. The snare will be localized over an average of cues from the two systems, and with attacks that are "smeared" by the 10-ms delay.

The three-to-one (3:1) rule can be explained in several ways, but it basically states that for sources of equal loudness, the distance from a microphone to its intended source should be 1/3 (or less) of the distance to an adjacent source, or another microphone. This does not apply to stereo microphone techniques, as they should be treated as one stereo microphone system rather than two separate microphones that may be panned to the same position. Following this rule will help to manage leakage and preserve phase relationships between sources, although varying the polar pattern and the orientation of the microphones can reduce the need to strictly adhere to this practice. It is good to keep it in mind, however, when positioning both the instruments and the microphones.

Another application of the theory behind the 3:1 rule is as follows: When combining two systems (mono or stereo) at equal gain, the more distant system should be placed at least three times the distance to the source of the closer system (see Figure 6.4). This action will serve to avoid the unpleasant artifacts of a "comb filter", which exhibits a rather "phasey" or hollow sound. The effect is due to the difference in time of arrival of the source at each system, where some frequencies will sum and others will cancel, depending on the phase relationship [2]. In a plot of frequency vs. amplitude, the cancellation points will appear as a series of dips in the resulting graph, resembling the teeth of a comb (see Figure 6.5).

This small ab/big AB technique can yield a very impressive overall sound, with power and bloom, yet in my experience it will require the addition of more support microphones at higher levels in the mix for focusing details, as in the case of the snare-drum solo. The two systems may

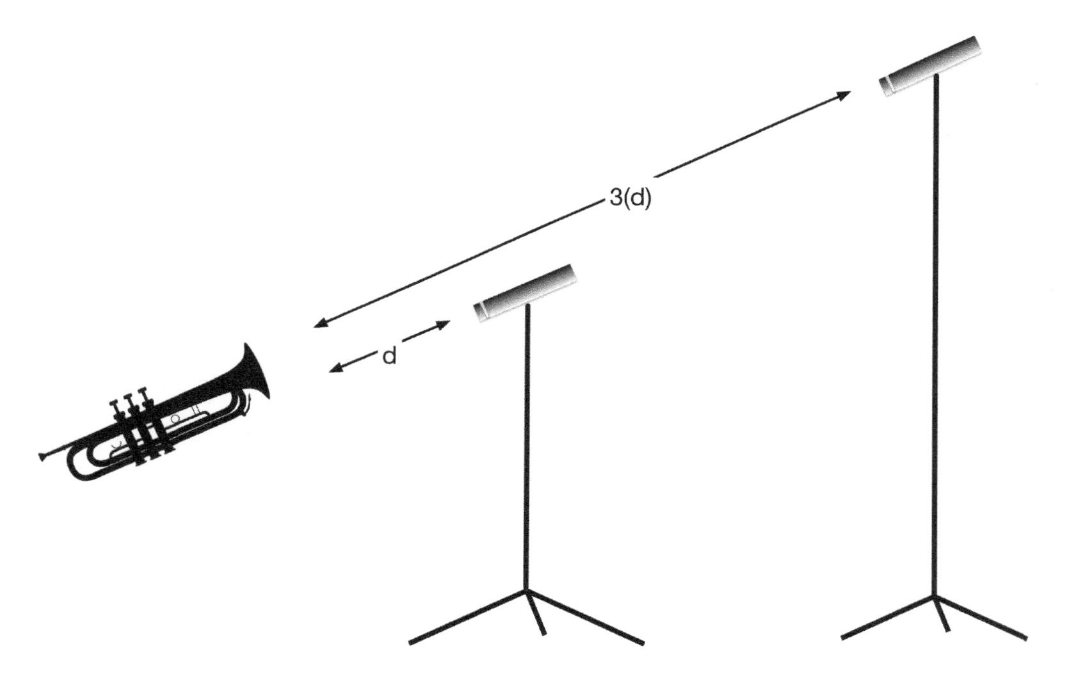

Figure 6.4 Diagram of a Source Captured with Microphones at Different Distances

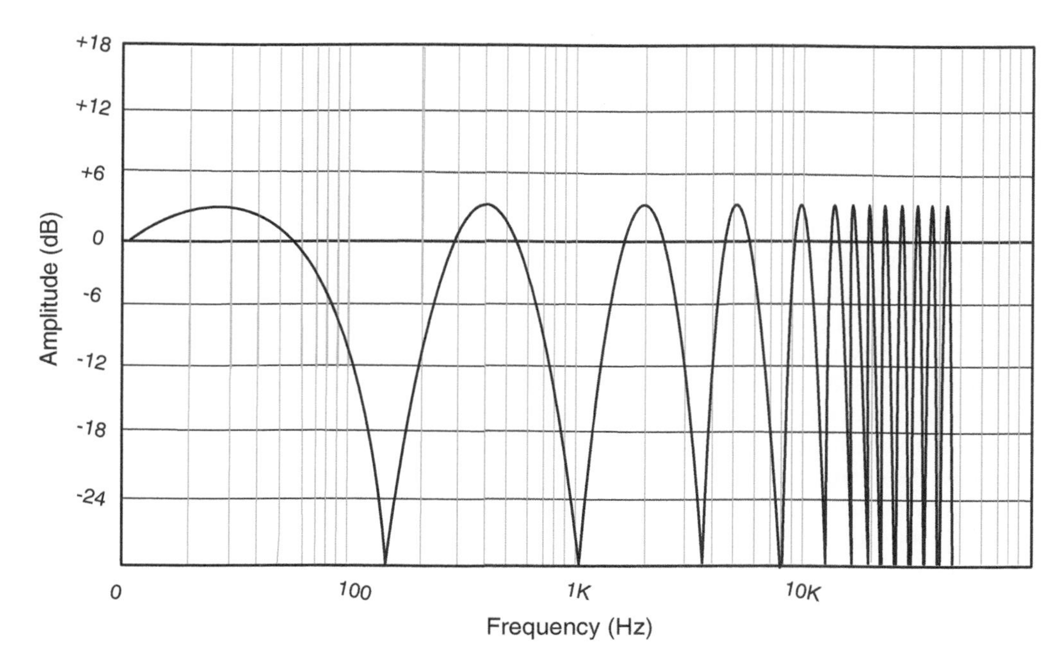

Figure 6.5 Theoretical Diagram of a Comb Filter

also be "time-aligned", where the closer pair in the previously described case would be delayed 10 milliseconds (ms) so that it is more or less synchronous in time with the more distant pair. This is a very complex undertaking, and even when exact measurements are taken of the microphone positions, the delay setting must still be "tuned by ear", since there will be timbral artifacts caused by the combination of the two pairs. Simply adjusting by 1 or 2 ms can change the overall timbral response, as certain frequencies are attenuated, and others are supported when the two systems are combined. For a greater explanation of applying delay compensation, see Exercise 10.1.

6.2 Microphone Placement

As the main microphone system comprises anywhere from 50–100 percent of the balance and sonic characteristics of an orchestral recording, its configuration and placement are absolutely critical. All previous descriptions of microphone techniques are meaningless unless the microphones are positioned properly and on the basis of some informed opinion.. Trying to record from that mystical "best seat in the house" will fail miserably if it happens to be located halfway back in the house (or farther). Even the highest-quality microphones lack the intelligence and discretion of the auditory system of most humans and therefore cannot pick out the subtle details of certain instruments over others at a large distance from the source, where the placement is most likely very deep in the late reverberation field. A better starting point would be somewhere above and slightly behind the conductor's podium, where the overall balance of the orchestra should be more or less even.

The best way to present a powerful and full-bodied string sound is to make sure the main microphone system is optimized for the capture of the string section. Other sections can be brought into focus with support microphones, but if the strings are not properly presented in the mains, it will be very difficult to successfully balance them with the rest of the orchestra through the use of spot microphones.

Beginning from where the conductor is listening should provide an overall sonic picture that closely represents what is being assessed on the podium. Of course, there may be a noticeable lack of detail from a few instruments, but the idea is to begin with a general soundstage that can be refined in a few quick microphone moves, after some informative listening. If we look back to Chapter 2, we will be reminded that the most important attributes to consider are timbre, balance, localization of sources, image and the ratio of direct-to-reverberant sound. Since all of these parameters are interrelated, they must be considered all together in one listening session in order to decide how the microphone placement might be optimized. For instance, moving back toward the hall might help correct a direct-to-reverberant balance that is too dry, while upsetting the overall timbre, as some general brightness or clarity is lost. Some issues such as overall timbre might be addressed more successfully by changing the microphone type, as certain microphones are known to be brighter than others. The following section addresses these main attributes and offers possible solutions for correcting each of them.

Adjusting for overall frequency response: Generally speaking, timbre can be adjusted by moving the microphones closer in for a brighter sound, and farther away for a darker presentation. What is important to note is that this adjustment can be made vertically by going lower and higher, or horizontally by moving forward and backward, or a combination of the two. Generally speaking, vertical adjustments will result in less change in the overall balance of strings to winds and brass, whereas a horizontal move back into the house may result in the brass and winds becoming more distant than the strings, thereby upsetting or perhaps improving a previous balance.

Direct-to-reverberant ratio: In order to get the right balance of direct sound to hall sound, it is best to start by adjusting vertically before moving front to back. It may seem more logical to move the microphones toward the hall to get more "hall sound" in the main pickup, as expected; however, many other attributes will be affected at the same time. Much of the string sound is directed in a vertical sense, which allows for some consistency in timbre when adjusting the microphone heights to correct for the amount of reverberation in the main system.

Perspective: This attribute is most quickly adjusted in a combination move that results in the microphones traveling on an axis that is at 45° to the stage floor. In other words, to get a closer perspective, one might position the microphones lower in height and at the same time farther forward toward the ensemble, by the same amounts. At a certain point, however, moving in too far forward over the conductor will result in the winds and brass sounding too "close" and quite possibly stronger than the strings.

Image and width: The desired image size and width is determined by the proximity of the main microphone system to the source, in combination with the spacing of the left and right microphones. For a single AB, the distance between the pair can be adjusted. A good starting place is from 1.2–1.5 meters wide, or 4–5 ft. Smaller distances between the two microphones will result in reduced spaciousness, whereas wider placements will begin to exhibit a reduced center image or a "hole in the middle" or the soundstage. When using two systems combined (close and far), the width of the image is affected by the balance of the two pairs. In the case of a Decca tree, the amount of center microphone in the mix greatly affects the width, as does the spacing between the left and right microphones.

6.3 The Secret of Recording

This was actually the short-lived original title of this book. The secret of recording – especially where classical music is concerned – is simply getting the microphones in the right place. For all recording engineers these are words to live by, and all other considerations are dependent on this very important first step. If this one basic task is performed successfully, the main microphones can easily account for more than 75 percent of the characteristics of the entire mix. In fact, the real secret of recording is to use the lowest number of microphones possible and paying special attention to their placement. While a purist might mandate a very small number of channels, a realist would strive to use a minimum number, even if that number is quite large. While the main microphone system should be the focus and priority in achieving a good balance and overall presentation, it may in fact require many microphones in combination with a main pickup to get the best final result, especially in a less-than-perfect recording venue.

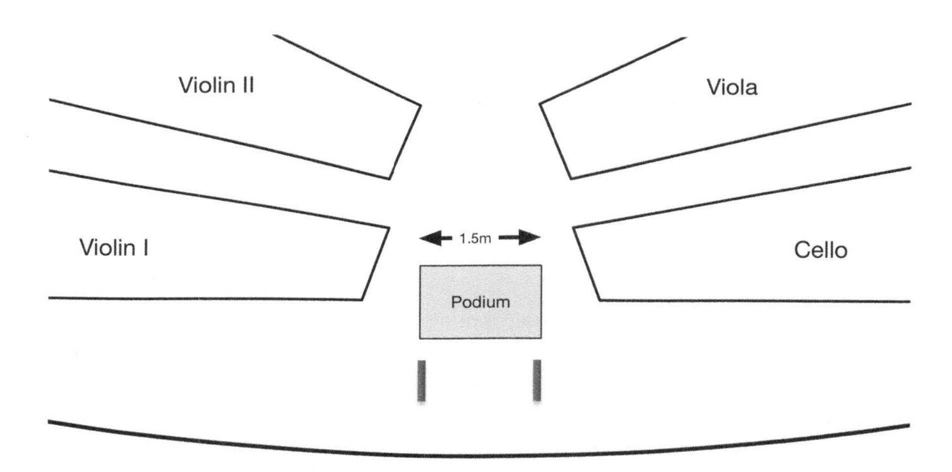

Figure 6.6 Diagram of the Main Microphone Placement in Relation to the Conductor's Podium

EXERCISE 6.1 LISTENING TO THE MAIN MICROPHONES

This can be done during a rehearsal or soundcheck. Most conductor podiums are around 1–1.5 m wide, and as such they provide a rather useful reference point or departure for placing the main microphone system:

1. Start off by placing your two omnidirectional microphones on separate tall stands and position them at the rear corners of the conductor's podium (the corners closest to the audience), or even a small distance behind, so as not to crowd the conductor (see Figure 6.6). Set the microphones so they are facing directly forward and parallel to one another, then angle them down about 30°–45°. Now set the microphone heights to

around 3 m (10 ft.) for a somewhat reverberant room, or 3.6 m (12 ft.) for a less live space. If you are worried about the musicians tripping on the stand legs, tape them down before anyone is onstage and walking around near the microphones. Make sure the conductor is aware that you've installed the microphones as well, and that they are comfortable with the placement.

2. Have a look at the resulting angle of the microphones in relation to the ensemble. The idea is that they are "looking" into the ensemble rather than over it, around halfway back. This beginning setting is rather arbitrary, as some microphones will perform better if they are angled higher up, and in some cases it might be preferable to have the main microphones angled lower into the string section for additional presence. As discussed in Chapter 3, a "perfect" omnidirectional microphone should have the same response on and off axis at all frequencies, but in reality, all omnis become somewhat directional in the high-frequency range, and this effect is more obvious in some microphones than in others. The microphones might have to be raised and lowered a few times until the angles are properly set and the two microphones look the same.
3. Set the starting gain of the microphone preamp to around 30 dB, and you are ready to listen – bearing in mind that the gain will need to be adjusted based on the microphone placement and the repertoire the orchestra is playing. This 30-dB setting is simply a rough starting point to ensure that signal is at least "present" when checking the lines. Note that if the microphones are lowered in height and brought in closer to the source, the preamp gain may need to be reduced. Also, it should be noted that a Mozart violin concerto would produce significantly less energy than a Tchaikovsky symphony.

A helpful trick for quickly returning a microphone to an exact height is to mark the microphone cable with a small piece of tape at the point where it touches the ground (or some other reference point on the stand). This will allow for a quick reset of the heights without having to measure or ask a colleague to check the heights of the two microphones from a distance. The "cowboy way" would be to forego the tape and simply stand on the cable at the exact point where it touches the ground. This is not recommended as standard practice in audio engineering, but it works in a pinch when time is criti-al and a roll of marking tape is out of reach. One must remember to keep the foot on the cable until the microphone has been returned to its original height.

As discussed in Chapter 4, it is commonplace in classical music recording to speak of the left-to-right perspective, as if the listener is looking at the stage from the audience. The main microphones, therefore, should be connected and recorded so that the left microphone is over the conductor's left shoulder, and so on, and this should be verified before the microphones are raised up to their starting heights. A "scratch test" should be performed before listening, to guarantee that left and right are connected properly (see Section 14.3 for more details). This may sound obvious and elementary, but it needs to be mentioned as it is incredibly important that all the lines are checked in this way before every recording session.

Adjustments should be guided by the parameters discussed in Section 6.2. At first listening, the best place to start is by evaluating the vertical placement on its own, trying higher or lower positions for the main system without changing the distance from the podium, to optimize the direct-to-reverberant ratio, timbre and overall perspective. This approach simplifies the process, and for the newcomer to acoustic recording of large ensembles, it is helpful to eliminate certain variables in order to achieve straightforward results. Take notes on the various changes in heights as the perceived differences can be very informative, and take a few starting notes on the positive attributes of the starting position before any changes are made. Also, if you think you might want to return to the starting position (you never know), it's always good to mark the height and floor position of the microphone stands. If time allows, small changes are best, in the order of 10 cm or 4 in. for each "move". This way, the optimal placement can be refined over several incremental changes, where each new position is based on the previous placement. Making smaller changes allows the listener to keep their frame of reference intact. In this way, the engineer will find that each new position sounds somewhat familiar or at least "as expected", except for the slight difference brought about by the height change, which then becomes the focus of the evaluation.

In the fairy tale of Hansel and Gretel, the children left a trail of breadcrumbs on their way into the woods so that they might eventually find their way out. Unfortunately, birds ate the crumbs, so they became lost. The audio equivalent of breadcrumbs is to keep track of all the significant changes, marking the height and the stage position of the microphone stand, so that it is possible to backtrack to a previous milestone at any point in time – for example, once the microphone position begins to sound as if it is too distant from the source, it can be easily and precisely returned to a previous lower or closer position.

During the height-adjustment process, other attributes will surely have been observed, for instance image width, or general balance of the sections of the orchestra. Now is the time to address these concerns, in conjunction with the previous set of parameters. The spacing between the main microphones can be adjusted to increase or decrease the size of the image, and this is very easily done if the main pair are on separate stands rather than a single bar mounted on one stand. Remember that using omnidirectional microphones in a stereo pair obliges the user to keep the signals assigned fully to the left and right channels, as the signal received at each microphone is more or less the same except for the difference in time of arrival. A narrower image, therefore, is realized by physically moving the microphones closer together rather than relying on pan pots in the mixer to reduce the image size.

General balance issues might be improved by moving the main pair up and over the ensemble to help focus woodwinds and brass, or down and out toward the house in order to give preference to the strings over the back half of the orchestra (see Figure 6.7).

There are many factors at play here, including timbral changes as well as the effect of the stage walls, ceiling design and possibly the proscenium, if there is one in the hall's structure. For example, heading up and over the string section may produce a brighter string sound while also decreasing the distance between the winds and brass and the main system, increasing the focus on the rear of the orchestra. Coming down and away from the winds and brass might actually result in a more direct brass sound into the mains, albeit more distant or delayed than the string sound, and a darker string pickup overall.

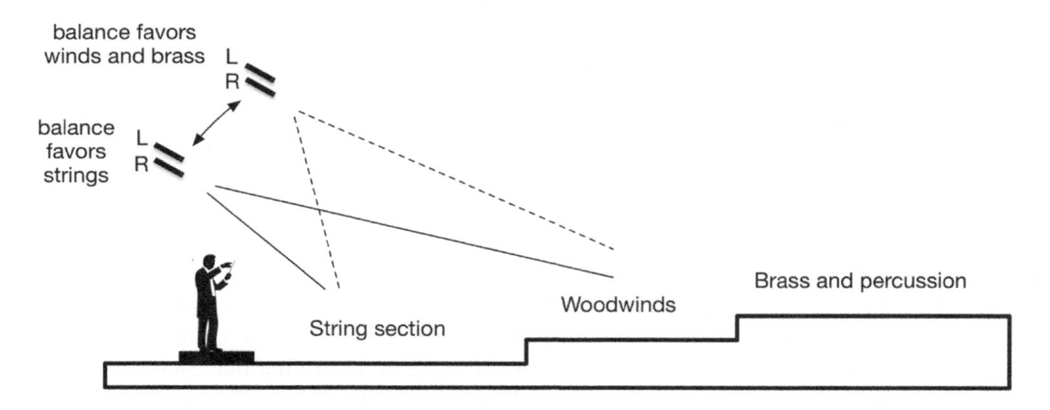

Figure 6.7 Diagram of the Main System in Relation to the Strings vs. the Back Half of the Orchestra

Much of this refinement is hit-and-miss, and decisions on microphone positions are made based on a combination of previous experience, logic and intuition. Locating a "sweet spot", or what is considered to be the best placement for the main microphone system, is really the secret of recording an orchestra, and the experience will be different in every hall. When using a close and distant system together such as small ab/big AB, some of the parameters from Section 6.2 will be evaluated and consequently adjusted at the mixing console. Through the combining of the two systems at various amounts, attributes such as image width, and to a point, timbre and perspective, can be manipulated without moving the microphones. The heights of each system, however, will still need to be addressed, and to a lesser degree, the positions of the systems in relation to the ensemble.

It is very helpful to record a rehearsal or soundcheck, and to place markers in the recording at times when changes were made on stage, to aid in coming to certain conclusions during later comparisons. Various main microphone positions can be reviewed at this point, and may help greatly in deciding on the optimal and final placement. If you are short of time during the rehearsal, playing back the recording afterwards can confirm that all microphones are working properly, and the individual channel gains are set correctly. In the case of a live recording, a playback can help to get a good mix ready before the start of the concert. In some cases, an orchestra will not allow a rehearsal to be recorded, so it is important to ask about the rules ahead of time. Normally it is possible that a rehearsal can be listened to over the microphones, even when recording is not permitted.

6.4 Room or Ambience Microphones

I have a rather strong opinion about room microphones, and I am sure that many very successful engineers would disagree with me on this count, but I find that room microphones are not as useful as one might think. Certainly, they are needed in live recordings for capturing applause and audience reaction, and for use in the rear channels of a multichannel/immersive recording. While I often set up a pair of room microphones, I normally leave them muted during a stereo mix, and here is why:

- To really find some interesting and properly de-correlated room sound that will enhance the overall combination of main and support microphones, it takes hours of experimenting to optimize and refine the placement of a room system.
- In a substandard hall, room microphones may risk bringing greater attention to the undesirable characteristics of the space.
- In an exceptional hall, it should be possible to achieve the perfect balance of direct and diffuse sound at the main microphones, as long as they are properly positioned.

I am a strong advocate for the use of artificial reverberation over room microphones, which will be discussed in more detail in Chapter 24 ("Mixing"). A good reverb program will offer complete control of reflections, diffusion, timing (pre-delay), size, shape, timbre and of course decay time. Most of these parameters are unavailable for modification when using room microphones alone. Improper use of reverb, however, can completely ruin a recording – great care should be exercised when using artificial reverb so that it doesn't *sound* like artificial reverb.

6.5 Chapter Summary

- Several equally valid techniques are commonly in use for main microphone systems.
- It is through general preference that the recording team chooses the technique they wish to implement.
- The "secret of recording" can be found in this chapter! Where orchestral recording is concerned, the position of the main microphones is absolutely the most important aspect of the recording process.
- In most instances, moving a pair of main microphones a few inches or even 10 cm will result in a pronounced difference in sound and balance.

References

1. Polymath Perspective. (n.d.). "The Decca Sound: Secrets of the Engineers," accessed June 11, 2023. http://www.polymathperspective.com/?p=2484.
2. Everest, F.A., and Pohlmann, K.C. (2009). *Master Handbook of Acoustics*, 5th ed. New York: McGraw-Hill.

Chapter 7

Recording Strings

Once the main microphones have been placed for an optimal overall balance of the orchestra in the room, the next step is optimizing the string sound and balancing the strings to the rest of the orchestra (winds, brass, percussion). This chapter will discuss the placement and adjustment of a widely spaced pair of microphones, commonly referred to as "outriggers", and address the best use of "spot" or support microphones for each string section.

Other sections of the orchestra can be supported in level and brought into focus using close or "spot" microphones, but the strings are not so easily accommodated. They are the closest section to the main microphones, which is a great advantage, but that alone isn't always enough. Other sections, including the woodwinds and brass, are generally projecting their sound toward the conductor and the hall, the main microphone system, and directly into and over the string section as a whole. Unlike the single snare drum from Chapter 6, the strings cover a vast area, normally half the surface of the entire stage. Therefore, in order for the spots to properly "cover" or represent each of the string sections, the microphones need to be set at a decent height or they need to be many in number. An acceptably close placement of one or two microphones per section should offer good coverage and a fairly even frequency response, but as soon as the rest of the orchestra starts to build its intensity, the strings risk being swallowed up. Typically, placing many microphones in close proximity to the instruments in each section will result in an aggressive and uneven pickup, requiring a great deal of processing to create a natural and blended string sound, not to mention more equipment and setup time. In either case, the spot microphones for the strings will suffer from having a great deal of brass and wind "leakage", and at times the brass may be even louder in the string spots than the strings themselves, depending on the repertoire.

7.1 Outriggers or Wide Pair

Making sure there is adequate string sound in the main pickup is incredibly important. Beyond that, an effective system should be in place to help the strings as necessary. It is good practice to find a method for overemphasizing the strings in general and then to reduce this effect slightly until a suitable balance is achieved. As mentioned in the opening of this chapter, sections such as winds and brass can be easily helped in a natural way using a few spot microphones, and as such the orchestra balance in the main pickup might even favor the string section, so that the string spots take on a less important function.

A widely spaced pair of microphones at the front of the orchestra can fulfill this role quite well. Known as "outriggers", like the pontoons found attached to a Polynesian canoe, the wider pair complements the main system by adding balance, power and increased low-frequency directivity [1].

DOI: 10.4324/9781003319429-9

The members of the Decca recording team were most likely the first to use the name outrigger in the audio world, as they commonly added this wide pair to their tree configuration, resulting in a five-microphone array. Producer Steven Epstein and engineer Bud Graham at Columbia Masterworks/Sony Classical referred to this additional wide pair as "pins", since they were used to "pin down" the outer string sections of the orchestra.

The outriggers are normally placed in line with the main microphones, at around the third desk of strings or halfway between the mains and the edge of the ensemble (see Figure 7.1). The most common polar patterns used are omnidirectional or wide cardioid, for extended coverage of the string section. Using an omni pattern will also increase the amount of hall sound in the balance, which is why a wide cardioid might be substituted depending on the quality of the recording space or the desired amount of control over the natural reverberation.

Some engineers prefer to use a multipattern microphone for the outriggers, so that they can adjust between omni and cardioid polar patterns. When using microphones with a power supply, the polar pattern selection can be easily changed at the floor without having to lower the microphones to reach the selector switch.

In either case, there will be a fair amount of signal from other instruments in the wide pickup, but this can be managed by the placement. For instance, lowering the height will yield more strings in the outrigger pickup as compared to other sections of the orchestra. Since the microphones are roughly the same distance from the source as the main system, the combination is mostly coherent. In fact, any "error" in phase correlation might be considered an additional benefit as it positively affects string blend.

The Decca team generally set the outriggers at the same height as the mains; however, a lower height setting of the wide pair will allow for more detail in string sound, thereby changing the principal functionality from "wide mains" to "omni string spots". A lower height also means that less level is needed from the outriggers in the mix to focus the strings, leaving the mains to dominate as the principal pickup. The use of mains and outriggers as a two-system array means a great deal of flexibility is achieved, as these can be refined as needed during a careful rebalancing session. Increasing outriggers in the mix means wider overall image and a better blend in the string sound, as the rear players of each section are supported more than the inner desks of strings, which exhibit more presence in the mains.

Microphone model and type for the wide pair should be of a smooth frequency response, and one that is forgiving rather than aggressive or grainy when placed in close proximity to the violin section. Equalization can help with this issue, but any change to improve the strings may work against the brass sound that is also present in the outriggers. Most likely a static balance will be decided upon between the mains and outriggers as they work together to establish the width of the image and the overall blend. An omni with a spherical attachment may be used in an effort to emulate the M50 as used in the Decca outriggers, but the engineer should be aware of the rising high-frequency response or general presence boost on axis caused by the attachments. My personal feeling is that this effect is quite the opposite of what is required from the wide pair, especially when they are placed at a lower height than the main system. An audio comparison of a main system and outriggers can be found on the companion website (https://routledgetextbooks.com/textbooks/9781138854543).

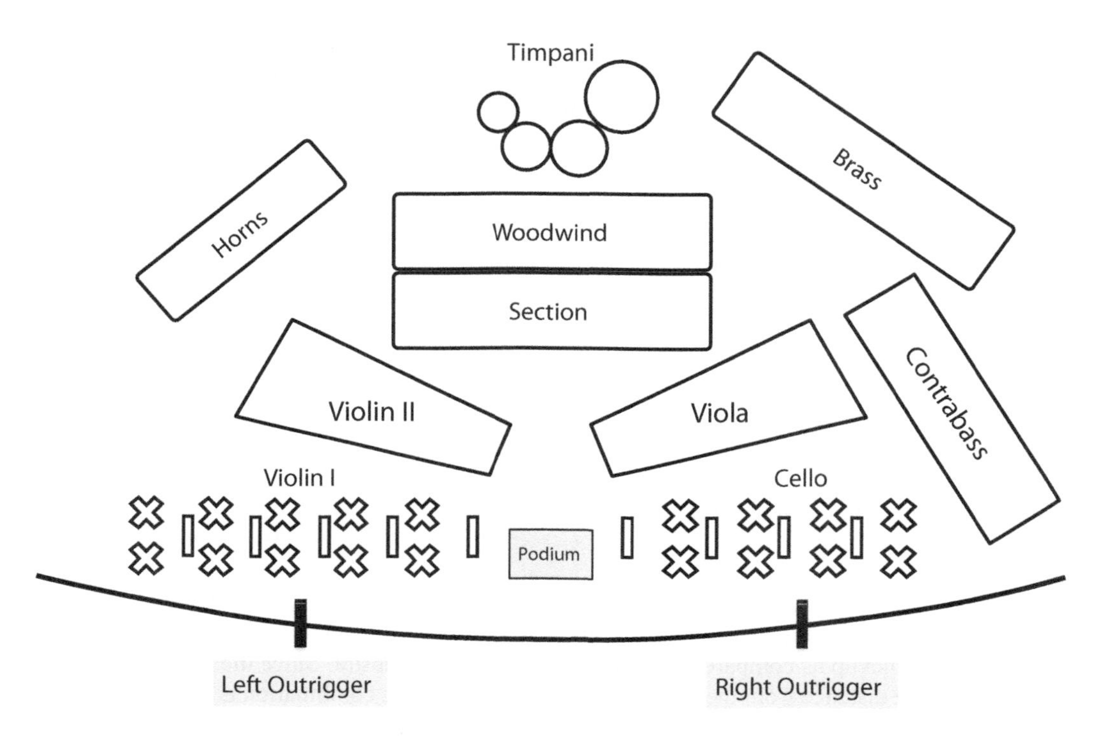

Figure 7.1 Diagram of Outrigger Placement with Orchestra

EXERCISE 7.1 PLACING AND EVALUATING THE WIDE PAIR OR OUTRIGGERS

1. Depending on the size of the string sections, set each of the microphone stands just behind the chair legs of the third stand of strings on each side (presumably the violins on the left, and either the cellos or violas on the right), as in Figure 7.1. Align the microphone stands with those of the main system or place them slightly closer to the orchestra. Set the heights at around 2.4 m (or 8 ft.) or slightly higher. Face the microphones toward the rear wall of the stage and set the vertical angle to around 45°, focusing the pickup directly into the closest row of strings. This way, the resulting presence capture is biased toward the strings rather than the instruments farther back onstage. Once again, if you are worried about the musicians tripping on the stand legs, tape them down before anyone is onstage walking around near the microphones. Make sure the conductor or stage manager is aware that you have installed the microphones and that there are no concerns with the placement.
2. Back in the control room, while listening to the main microphone system, bring the outriggers into the mix. Exaggerate the amount at first, going past the level of the mains, so that it is obvious how the outriggers affect the overall image and the string sound.
3. Settle on a preferred amount of outrigger signal in the mix, based on how much help or support the strings need in the balance, and how much the wide pair is needed for good

section blend and image width. The level will usually be somewhere between -15d B and 0 dB if the input gains are closely matched to the main system. Experiment with muting and unmuting the outriggers while listening, to observe their role in the mix. It is also good to listen to the wide pair on their own, at the prescribed level in the mix, to quantify how much of their signal is needed to improve or enhance the main microphone system. This process can aid in refining the positioning of the mains and outriggers, or in determining a good starting position for the microphones in future projects.

In certain cases, a strong and well-blended string section along with an excellent hall may render the outriggers unnecessary – and in these cases the engineer should feel free to simply leave the outriggers out of the mix.

7.2 String Spots

Clearly, the best way to support the strings in a general way is through the use of outriggers, as described in the previous section. Using omnidirectional microphones is good practice because of the lack of off-axis coloration, so that all instruments in the area near the system are captured naturally. That being said, spots for each string section are still quite often needed for coverage of solo passages, or for moments when the brass and percussion are completely obliterating the strings and some bite is needed just to maintain that they are still part of the group. In this case, simply placing a microphone on each of the first desks may be enough, perhaps enhanced with an additional microphone halfway back in each section. Popular music presentations involving rhythm section performing together with orchestra will most likely require one spot microphone on each stand of string players, or a vast number of miniature microphones that can be attached to each instrument in order to balance the strings.

First violin: the concertmaster will frequently have solos in larger works, so it is good to have a microphone ready to go that covers the front desk players. Positioning the stand can be tricky, since the players need elbow room for bowing, the conductor needs ample room for baton-waving, and it is nice not to have a stand directly between the concertmaster and the audience. A very helpful technique in placing string spots is to think about sound sources on the back of each microphone, in order to increase the odds that the recorded signal will be useable. With all of these considerations in mind, the best possible spot for the microphone stand might be just behind the assistant concertmaster, next to the left rear leg of their chair. The microphone can then be brought over the two players through the use of a boom arm on the stand, and angled toward the principal player, but also looking down the section, and mostly away from the brass, at a height of around 2 m, or a little over 6 ft. As with the other string sections, a second microphone may be added farther back in the section; however, at this point the source will include much more percussion, possibly horns or clarinets and so forth, so its usefulness will be reduced.

Second violin: the same position and height as for first violin works here as well – behind the assistant principal's chair, next to the left rear leg (from the player's perspective). This way the stand is out of the way of the conductor and the player's bowing arm, and the microphone can be positioned to favor the front desks with an angle that allows for good capture through the section. One mostly overlooked use of a spot microphone on the second violins is to provide warmth to the whole violin sound. Since the second violins normally double first violins one octave lower, the orchestration can be made to sound richer by helping the bottom octave of the violin lines.

Viola: viola tends to need more help with presence, as it is generally a darker instrument than the violin. This requires a lower position, which helps for definition and control of leakage, but also reduces coverage. This is typically not an issue, as the viola section normally has lower numbers than the violins, so about the same coverage is achieved at the lower position, around 1.5 m or 5 ft. Covering the section from the front tends to leave the brass section directly on axis if the viola section happens to occupy the inside right stage position. In that case, it makes sense to go in somewhere behind the first stand and orient the capsule forward, so it is looking away from the brass section. The resulting pickup is not ideal for the viola but at least it will be useable as viola capture, and helpful for covering solos.

Cello: when the cellos are seated on the outside of the string section close to the edge of the stage, their sound will be well supported and enhanced by the right-hand outrigger. However, solo coverage is still an issue, so the placement of a spot near the principal player is recommended. With the head of the cello neck extending over the player's left shoulder and because the bow activity is much lower down than the violin, the right rear leg of the chair is a better reference point for the placement of the microphone stand, and again the assistant principal's chair is a good spot since it is inside the ensemble and less distracting to the conductor (and the audience in a live setting). The microphone height is generally lower than for the other string sections, at around 1.4 m, or a little under 5 ft.

Contrabass, or double bass: another generally positive attribute of the wide pair is that they tend to capture a "silver" or rich quality in the high-frequency range of the double basses, which I describe as a pleasant "V" sound off the bow. Combined with this effect, it is always good to have one microphone positioned about 60 cm or 2 ft. off the floor and very close to the principal bassist, as close as possible and just under the height of the bridge so as to be out of range of the bow's path. In this way, if there is any lack of double bass sound or general low frequency in the balance, it can be easily fixed using a strong signal, with very little leakage from other sections due to the close placement relative to the source.

A condenser microphone with a large diaphragm is a good option for the principal bass, although a small diaphragm condenser can work just as well, except for a possible increase in directionality and therefore less leakage. Many engineers place a bass-section microphone very high up, similar to the violin sections, but this configuration practically mirrors the capture of the basses in the outriggers (or in the adjacent string spots for that matter). Unlike the other string section spots, it may be more common for the double bass channel to be "built in" to the mix at a static level, for constant reinforcement of low-frequency content and detail.

Observing the orientation of a microphone's "null" side is a powerful tool in orchestral recording. When positioning any directional microphone, it is good to consider facing the rear of the microphone toward the most likely source of unwanted leakage.

7.3 Chapter Summary

- It can be very difficult to build the string section sound using spot microphones, so ample effort should be allocated toward optimizing the main system.
- "Outriggers", or a widely spaced pair of omni microphones, can help to enhance the overall string sound more effectively than closely positioned cardioid spots.
- The contrabass or double bass microphone is positioned differently and is treated separately than the other string spots in the mix.

Reference

1. Woszczyk, W. (1992). "Microphone Arrays Optimized for Music Recording." *Journal of the Audio Engineering Society*, 40 (11) November, 926–933.

Chapter 8

Recording Woodwinds, Brass and Percussion

While many sources divide the symphony orchestra into four sections, it can be said that the ensemble is really made up of two fairly equal halves – the string orchestra, occupying the front half of the stage area, and the woodwinds, brass and percussion. This second grouping represents the entire back half of the orchestra on stage. This chapter covers best practices for stage positions and capturing and balancing each section, including microphone choices and placement. Basic descriptions and traditional seating plans for each group are laid out in Chapter 4.

In an ideal recording situation, all sections of the orchestra should be perfectly balanced and in focus in the main pair, so that spot microphones are not needed. This is, unfortunately, a rare occurrence. Remembering that a main system comprised of omnidirectional microphones will yield an exaggerated sense of depth, it should be expected that the back half of the ensemble will need to be brought into focus through the use of support microphones.

8.1 Woodwinds

There are several approaches to capturing the woodwind section, specifically comprised of flute, oboe, clarinet and bassoon (and occasionally saxophone). All of these are valid and acceptable techniques, so it is simply up to the recording team to decide which is the best choice for each scenario. Readers are again encouraged to try all of the proposed techniques described here in order to form their own conclusions.

Spaced pair: this is the simplest approach, offering a two-microphone solution to bring the woodwinds into focus with good correlation (see Figure 8.1). While spaced cardioids offer good coverage, the issue of image can result in compromise. The ideal microphone position that covers the entire section is normally too wide to use as a fully assigned left/right element in the mix, and will need to be "panned in" so that the woodwind section has an appropriate image matching that of the main pickup. Since this system is secondary to the main microphones and will be much lower in level in the final balance, the impurity of the pickup will probably go unnoticed in the full mix. As with all the techniques which are placed in a single line, care must be taken to balance the first and second rows of the section. The microphones should be set so that they are on axis to the more distant second row of clarinets and bassoons, since the first row (flutes, oboes) is physically closer to the microphones. This technique offers little control over the individual instrument sections, so some experimentation will be necessary to optimize the placement. Several rehearsals in the same venue may be needed to be sure the woodwinds are well represented with a single pair of microphones.

ORTF: a near-coincident stereo technique is a good alternative here, as it can remain fully left/right in the mix, appropriately presenting the woodwind image to match the main pickup (see

DOI: 10.4324/9781003319429-10

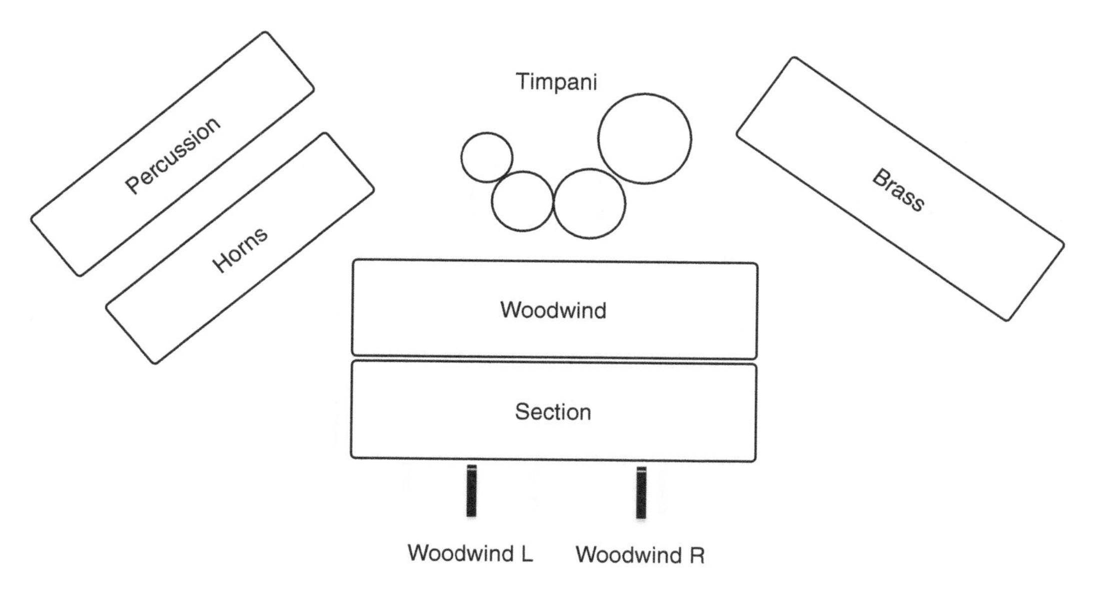

Figure 8.1 Diagram of Spaced Pair Woodwind Spots

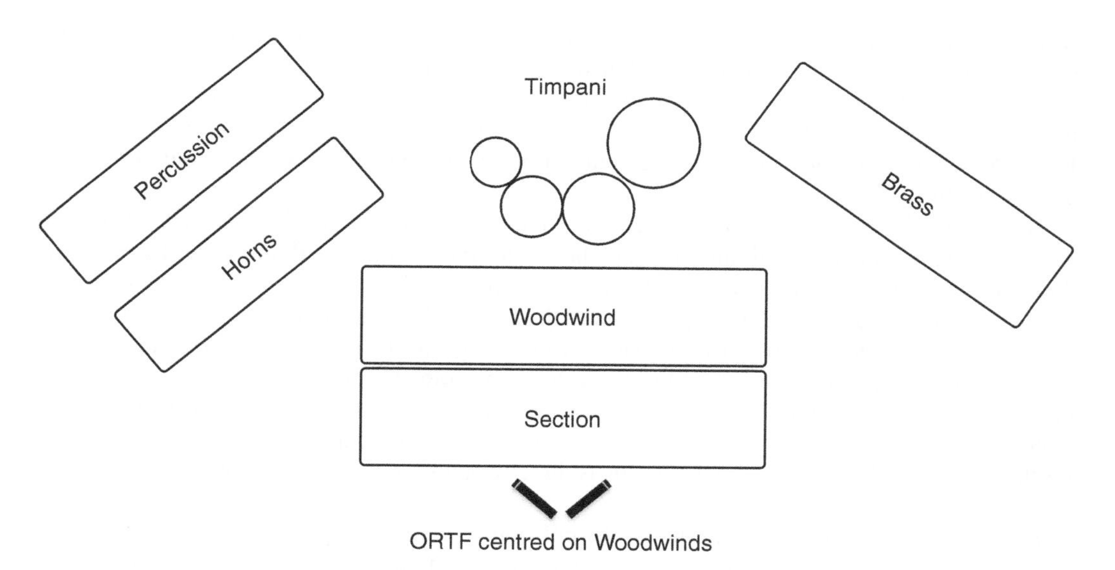

Figure 8.2 Diagram of ORTF Woodwind Spots

Figure 8.2). The practical advantage is that it is supported on one stand, so there is less hardware on stage. An added advantage of using a bona fide stereo microphone technique here is that the entire back half of the orchestra is evenly represented in the woodwind pickup, so that brass, timpani and percussion are all part of an intimate stereo capture that fits well with the main microphones.

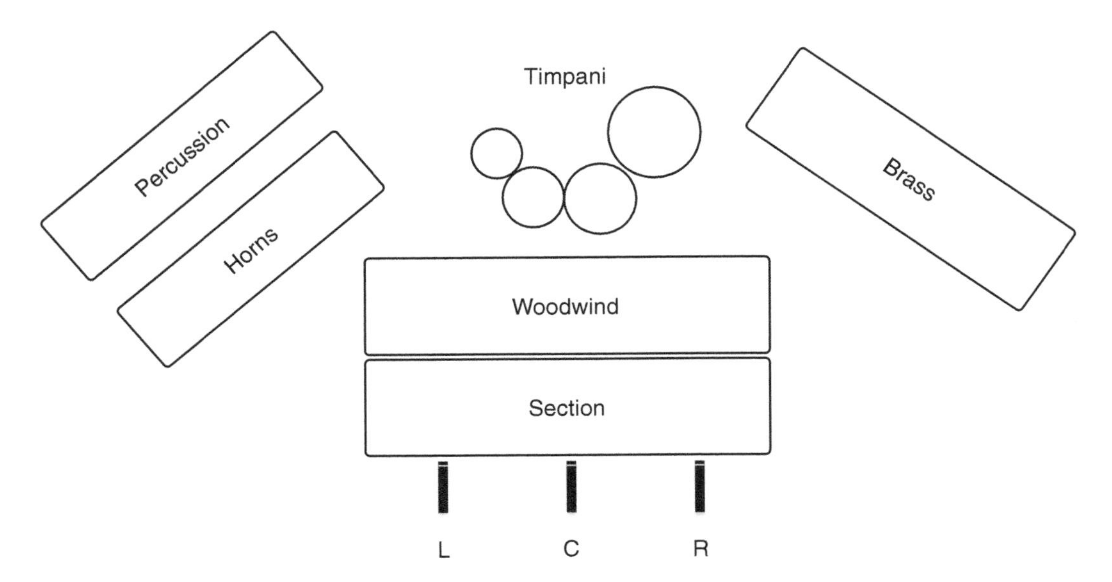

Figure 8.3 Diagram of L/C/R Woodwind Spots

In some cases, as long as the ratio of winds to brass in the woodwind pickup is suitable, an ORTF pair on the back half of the orchestra combined with the mains may be all that is needed.

Three spaced microphones: an alternative to using a spaced pair, three cardioids in a line can help to increase the width of coverage, while maintaining a natural image of the section (see Figure 8.3). For film scores and surround-sound projects, having an L/C/R woodwind pickup can be very helpful.

One microphone per instrument group: in commercial production, there is very little time afforded to experimenting with microphone placement or to working with the orchestra and conductor on internal woodwind balances during a recording session. The most common solution in this case is to place a cardioid microphone on each group within the section – separate flute, oboe, clarinet and bassoon microphones. For larger works, this can include microphones for English horn, bass clarinet, contrabassoon and in some cases, piccolo. The net result is a seven- or eight-microphone setup for complete control of woodwind balances (see Figure 8.4). Placement of each microphone stand is between the first and second player in each group. In this setup, microphones for the flutes and oboes will be angled to focus directly on the source, to attenuate the instruments in the second row.

Pan each instrument channel to match the position in the main pickup. More setup time and equipment are needed for this approach, including a recording system capable of accommodating many separate tracks, or at least a large mixing desk if the balance is being performed "live to two-track" with no multitrack recorder. As more spot microphones are deployed on the stage, the engineer will quickly discover that each spot will be introduced at a lower level in the final mix, as less level is required from each source to build an appropriate balance. As with the other microphones on stage, the positions will be adjusted based on listening to each resulting capture. For instance, a microphone set for an English horn player sitting near the brass may need a closer placement than normal.

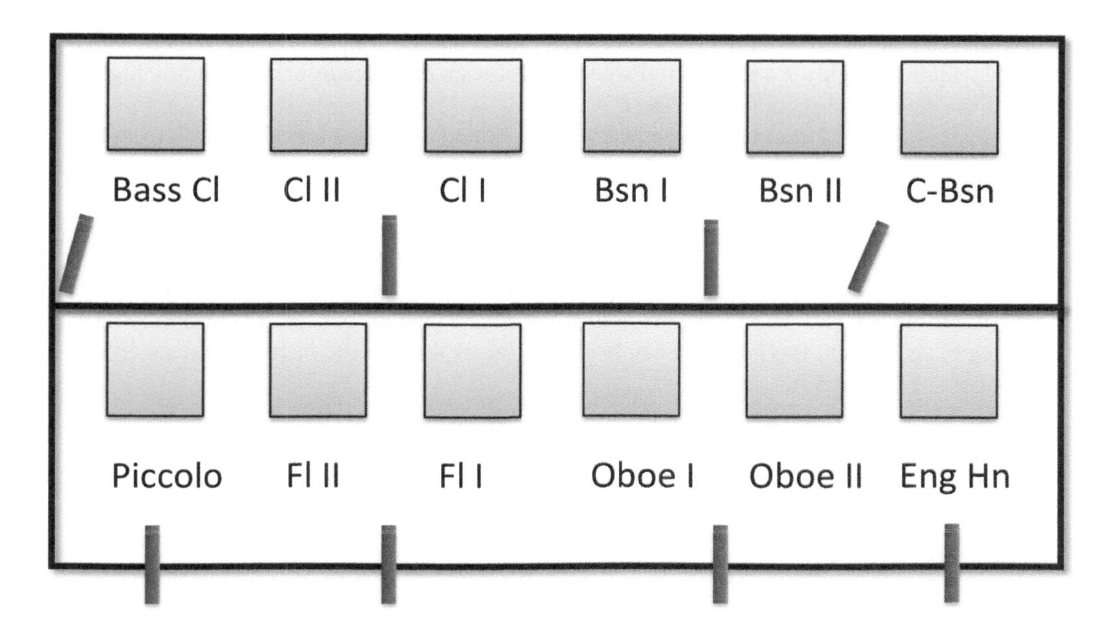

Figure 8.4 Diagram of Spots for Each Woodwind Instrument Group

Even in the most complex orchestrations and when massive orchestral forces are in play, a separate microphone for piccolo is probably not needed. This is an instrument that is rarely masked by any other; it cuts through easily because of its high range.

Bidirectional microphones: except for the ORTF option, any of the previous solutions may be carried out with bidirectional microphones. The advantage here is the use of the null side of the microphone to manage the leakage from adjacent instruments and sections. The microphones can be placed above the section looking down, so that the sides of the diaphragm face the brass and percussion, allowing for a more isolated woodwind pickup. The potential downside to using figure-8 microphones is that the polar pattern is narrower than for most cardioids, so it will be harder to achieve even coverage of the section.

EXERCISE 8.1 PLACING AND EVALUATING VARIOUS WOODWIND SPOTS

Try these different techniques for woodwind capture, all at once or at different times. Record each technique and compare:

1. Spaced pair – Left woodwind spot between Flute I and II, right spot between Oboe I and II. Height: roughly 2.4 m or 8 ft. if woodwinds are on the stage floor, and slightly higher if the woodwind section is on risers.
2. ORTF – Place a large stand on center between Flute I and Oboe I, slightly back from the winds. Starting height: between 2.5 and 2.7m (8.5–9 ft.).
3. One microphone per instrument group – heights will be lower for better isolation – flute, oboe, clarinet: around 2 m or 6.5 ft. Bassoon should be captured from the player's right side rather than above the instrument, at around 1.5 m or 5 ft. The sound coming from the hole at the top is mostly the reed, without the core of the tone. English horn may be slightly lower (1.5 m or 5 ft.) since it projects less than an oboe and is normally near the brass section. Bass clarinet can be fairly low as well, as the instrument rests on the floor – 1.5 m or 5 ft.

The best practice for refining all of these microphone positions is to assess the placements while the musicians are in their seats warming up before the beginning of the rehearsal or recording. Some musicians arrive very early, so this process can be completed over a time-frame of upwards of an hour before the session. Great care should be taken not to touch or bump into any instruments when moving around on stage while adjusting microphones or tidying up cable runs.

8.2 Brass and French Horns

The brass section functions quite differently from the woodwinds. Most orchestral writing employs a greater number of moving phrases through the woodwinds, whereas brass orchestration typically utilizes more of a chorale presentation, where blend is more of a priority than individual lines. A good brass section should be able balance itself with some instruction from the conductor. Aside from the occasional solo passage, it is more common that the entire section will need to be brought into focus rather than the individual elements.

Brass, excluding horns: microphone placement will depend on the seating arrangement. If the brass section is in one line, then two microphones will most likely suffice in covering the section and may be thought of as "Brass left and right". In the case of two rows of brass, the microphones will be specifically named Trumpet and Trombone. Unless the sections are unusually large, using one microphone to cover each instrument group while favoring the section leader should be feasible. For brass, it is best to place the microphones quite low and at a slight distance from the section for a warm, round sound. Tuba does not normally need a microphone, since its sound is normally well represented in the main pickup, and is better captured from a distance, not requiring much focus. If channel count allows, a separate tuba microphone may be added, or the trombone microphone could be named "Low Brass" and positioned to capture both the trombone section and the tuba.

French horn: while the horn is technically a brass instrument, its role in the orchestra is more similar to that of the woodwind section. With any luck it can be arranged so that the horns are

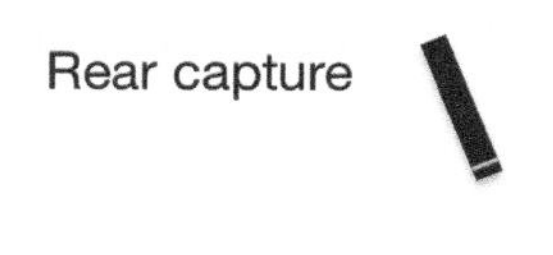

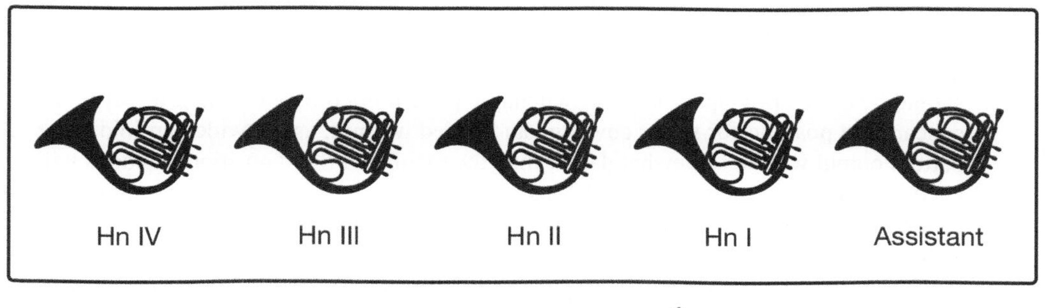

Figure 8.5 Diagram of French Horn Spot Placements

seated on the opposite side of the stage from the rest of the brass, in an effort to balance the energy from left to right and to create an interesting interplay of parts from each side. In most cases, one microphone can cover a section of four horns plus an assistant principal, but in the case of six or eight players, a second microphone will give better coverage. Two schools offer compelling and conflicting justification for microphone placement (see Figure 8.5):

1. Placing the microphone in front of the section will capture the horns as they are meant to sound, with a pure and rounded tone but without the spit and edge of the direct sound coming out of the bell. The resulting capture sound will be similar in sound to the main microphone system.
2. On the other hand, recording the section from behind the instruments results in a sound more typical of a support microphone, with more detail and attack than what is normally available from the front. The natural sound of the horn should be well represented in the main system.

Some engineers will place both a front and rear capture for the section, so both characteristics are recorded and can be used separately (as in either/or) or blended together.

An excellent option for the horn spot is to position the microphone somewhere directly above the section, where there is a combination of roundness of sound mixed with the clarity of the bell output, mixed with the added power of the floor reflections. This placement is most easily realized by hanging the microphone over the section, as the legs of a rather large stand will be disturbing to the musicians.

EXERCISE 8.2 PLACING THE BRASS AND HORN SPOTS

This is less of an evaluation than the previous exercises, but try comparing condensers and ribbons for use with the brass and investigating front and rear capture of French horns:

1. Trumpet and trombone: a warm-sounding large-diaphragm condenser microphone is the author's preference, although many seasoned engineers prefer ribbon microphones for brass. Place a condenser microphone about 1.5 m or 5 ft. in front of each section, and quite low in height (1.2 m or 4 ft., higher if brass are on a riser). Focus for the pickup should be on the principal players. Switchable pattern microphones might be set to the subcardioid position for wider coverage and added warmth, as the wide-cardioid polar pattern output will be somewhat darker than the regular cardioid position. Ribbons will need to be placed much closer to the instruments (1 m or 3 ft.), and will therefore not cover the section as well, possibly requiring the addition of extra microphones.
2. Tuba: any microphone which can offer a small amount of focus will be adequate – remember the tuba sound is 90 percent in the main microphone system, and the spot should only be needed to provide a "point" on the attack of each note. This may sound uninspired, but whatever microphone is still available at the end of the planning stage is probably fine for tuba.
3. Horns: again, a large diaphragm for warmth is best. If using the front position, try a height of 2 m or 6 ft., possibly higher, while bearing in mind that leakage may be an issue. For the rear placement, 1.2 m or 4 ft. high should avoid some of the direct sound from the bell while incorporating a good amount of floor reflections. In either case, some effort should be made to reflect the bell sound out toward the hall and the main microphone configuration – either a nearby stage wall, or some portable reflectors. Scrap plywood can work well, or even large road cases parked in behind with their lids open will help. Be careful not to crowd the musicians, as they might end up feeling uncomfortable and may not play out to the hall. Position the reflectors at least 1.5 m or 5 ft. back behind the bells to avoid this issue.

8.3 Percussion

The percussion section can be broken down into timpani, mallets (such as xylophone and glockenspiel), and the untuned instruments (such as snare drum, bass drum and cymbals). Typically positioned at the greatest distance from the main microphones, percussionists have learned to anticipate the resulting delay by playing slightly ahead of the beat at all times, so that their parts are heard in sync with the players at the front of the stage, and with a suitable volume. It is the engineer's task to make sure the various instruments are not too diffuse as is quite common for prominent sources coming from the rear of the stage. The goal is to provide focus from the spot microphones, without increasing the level.

Before the spot microphones are introduced, the clarity of percussion instruments can be improved through the use of harder mallets or sticks, as necessary. For the timpanist, their playing may sound perfectly concise with sharp attacks from where they are sitting, but by the time the sound of the drums travels out to the main microphones it will be less clear. A stick with very hard felt might in fact sound horrible to the player at the drums, but it might be perfectly balanced at the main pickup between attack and fundamental tone of the instrument. Improving the drum sound acoustically in the room is a much better solution than using a bright spot microphone signal to mask a muddy drum sound in the main system. Having a player listen to the differences in their mallets during a playback session can help to convince them of the importance of this issue. Xylophone, marimba and glockenspiel can be refined in this same way, by trying either a harder or softer mallet depending on what is needed.

Timpani: as the timpani is normally loud enough in the main pickup but quite often sounds "thick" or "boomy", it is best to start by speaking with the timpanist, as already mentioned. Short of that, small diaphragm condenser microphones are the best choice for spots. Generally, one microphone will suffice to add a suitable amount of attack to four drums in an even coverage. More than four drums or two players will require a second microphone for even coverage. Ideally the timpani would be positioned in the center behind the woodwinds, with the various percussion instruments spread out to either side.

Mallets, or tuned percussion: this includes chimes, xylophone, glockenspiel, marimba, vibraphone and crotales. Quite often the attack of each note is heard, but without much differentiation of pitch. The goal of the spots in this case is to tidy up a diffuse presentation (common to xylophone and "glock"), and also to give clarity to the pitch of each note. In a very large percussion setup, it is important to get at least the mallets within good placement of the spot microphones, possibly at a height of 2 m or 6.5 ft. from the floor. Chimes or tubular bells should be captured away from the heads of the chimes where they are struck.

Untuned percussion: this category includes snare drum, bass drum, cymbals, tam-tam and all the accessory percussion – tambourine, woodblock, triangle, guiro, bongos, wind machine and thunder sheets, and the list goes on. The best approach in a complex setup is trial and error – set out a number of spots and listen for what isn't properly covered, and then adjust from there. Certainly, finger cymbals and water gongs will need more help than the snare drum, although the snare risks having a very diffuse or "roomy" presentation if there isn't a microphone within a reasonable distance that can be used to "dry up" the sound when an amount of clarity is needed.

With orchestral bass drum, it can be difficult at times to get a good deep note and rhythmic detail. One approach that works well is a dynamic microphone set at only 10 cm or 4 in. from the head. This will provide good detail during soft passages, and substantial low-frequency energy in louder passages. Some compression may be necessary to manage gain on this channel, because of the close proximity to the drum.

Similar to the piccolo, as mentioned in Section 8.1, even the softest triangle passage will find its way into all the microphones and is therefore very difficult to control when played too loudly. A good analogy is the tiny red sock that mistakenly ends up in the washing machine with a large pile of white clothes – the entire laundry load turns pink.

EXERCISE 8.3 POSITIONING OF PERCUSSION INSTRUMENTS AND THEIR SPOTS

Once the musicians have settled on the placement of their instruments, bring in the spot microphones, covering the full section as needed:

1. Timpani: center, if possible, with small diaphragm condenser microphone over the drums. Check with the player that the microphone stand isn't blocking their view of the conductor. The goal is to capture point, since the deep tone of the drums will be present in the main capture system.
2. Other percussion: with the help of the conductor, the recording team (usually led by the producer) can ask whether certain instruments can be placed farther left or right to make a more interesting spread across the channels in the recording – for instance, crash cymbals one side, snare on the other, bass drum and tam-tam near the middle. Also, some mallets may be split – xylophone one side and chimes on the other. The actual split will depend upon who is playing what part: for instance if "Player 1" is assigned xylophone and chime parts, those instruments will most likely need to be close together, unless the player has time to move between the two spots during the performance. The players generally do not like splitting up the instruments, as it affects their ability to play together with accuracy. Generally speaking, after a short period of adjustment, the musicians will adapt and perform very well in a spread left/right configuration.
3. A good starting height for most of the percussion spots is around 2 m or 6.5 ft., and slightly higher for timpani, depending on the number of drums. For chimes or tubular bells, capturing the instrument from above will result in a recording of the nasty "click" produced as the rawhide or plastic hammer impacts the head of each chime. The more desirable and warmer sound of the resonating pipes is found much lower down, so the microphone should be set approximately 1.2 m or 4 ft. from the floor.

Where the percussion spot microphones are concerned, it is a good idea to choose a cardioid that exhibits an even off-axis response, as any one microphone in the percussion section will be capturing many different instruments, from all angles. Leakage will be a positive and desired attribute and will need to be pleasant-sounding and natural.

8.4 Chapter Summary

- Fewer woodwind microphones will require more time to balance; more microphones offer better control but take longer to install.
- Horns and brass should be separated left and right whenever possible to allow for a more impressive presentation.
- French horns are designed to be heard in a reflected rather than a direct way, so it should be ensured that their energy is reflected toward the mains, and that a good amount of reflected sound is included in the close pickup.
- In large percussion setups, be sure to have a good shot at the mallet instruments with the spot microphones.

Chapter 9

Recording Harp, Piano, Celeste and Organ

These instruments end up in many different positions within the orchestra – and their placement is usually determined by where they fit on stage; or, in the case of the organ, where it happens to be installed. This lack of standard placement can lead to issues with the sightlines between the various players and the conductor, the physical comfort of the musicians, and how well the instruments balance naturally with the rest of the orchestra. Piano and celeste are quite often categorized as part of the percussion section, although they aren't necessarily positioned near the percussion instruments on stage. Some helpful advice on how to deal with these instruments is presented in this chapter.

9.1 Harp

Throughout the vast catalog of orchestra recordings, harp balance tends to be overdone rather than "under-mixed". This is probably because it is very difficult to balance harp with orchestra, and because it appears less frequently in most orchestrations, exposing or over-mixing the harp for a few seconds here and there is an exciting effect. A favorite contrasting description comes to mind – "subtle, yet bold" – as the harp sound emerges from the orchestral soundscape and lands an exciting glissando on the ears of the listener, fulfilling the composer's wish for an unexpected wash of sparkle and texture. Harp may be placed in various positions on stage, as it really depends on where adequate floor space might be found. Quite often, the harp or harps will be placed on a riser if they are either set farther back behind the violin sections or tucked in to one side of the woodwinds. An ideal location is around halfway back on the left side, between the first and second violins, if there is adequate space (see Figure 9.1). This allows for good capture of the harp in the left outrigger, which will sound fuller and more natural than a closely placed cardioid spot.

In a recording session where a stage extension can be used, it may be advantageous to place the harp or harps out on the front deck, next to the violins and off the left shoulder of the conductor. Here they reside quite closely to the left outrigger, and the harp spot microphone will have less leakage from other instruments. Harpists generally like this placement, after a short period of adjustment, as they are close to the conductor, and they can hear themselves very well. Repertoire with two harps may benefit from a split placement, although the harpists will most likely prefer to sit next to each other for performance reasons.

Placement will not be ideal, but it is possible to capture two harps with one bidirectional microphone between the two instruments, if channel count is limited. Adjusting the physical placement of the microphone between the two instruments can help to refine the balance of the two instruments.

DOI: 10.4324/9781003319429-11

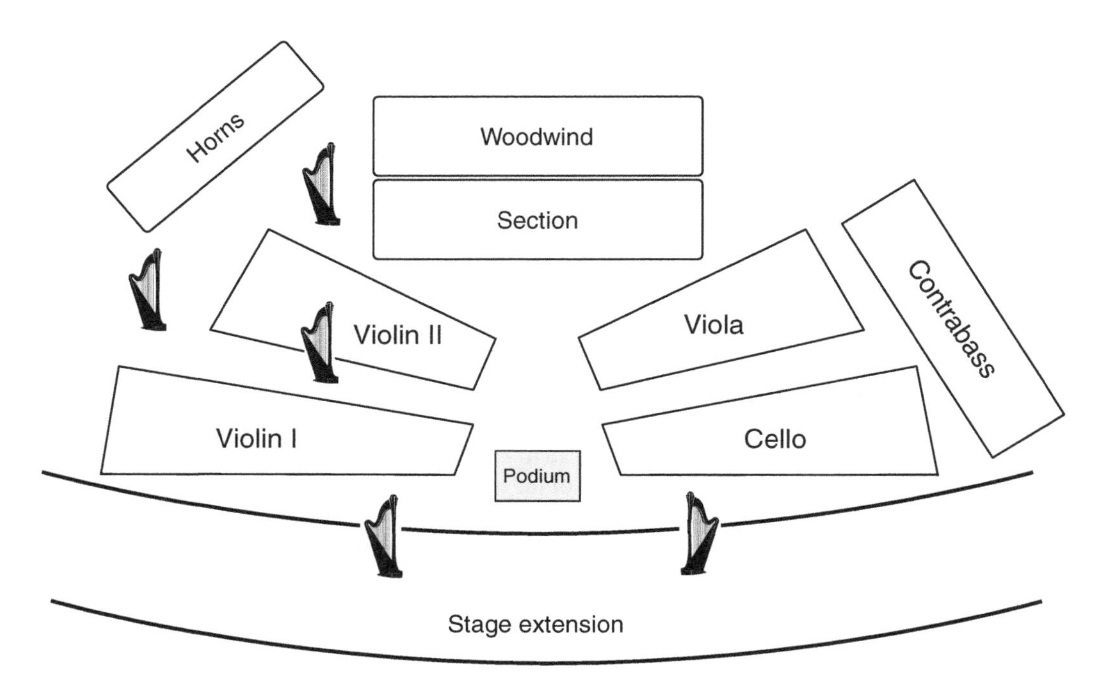

Figure 9.1 Options for Harp Placement Within the Orchestra, or on the Stage Extension

The hardest part of capturing harp is obtaining an even pickup from the low to high strings. The lower, warmer strings of the instrument frequently lack the ability to cut through the dense textures of the surrounding string sections, whereas the high strings tend to jump out with significantly more attack in the sound than actual pitch. The spot microphone should therefore be placed so that it favors the lower part of the instrument, facing the soundboard, which is where the core of the sound is produced (see Figure 9.2). Capturing the high notes off axis to the spot will help to smooth the often-piercing attacks of louder passages and the final notes of an ascending glissando.

Harpists will place the music stand to their left side, so the microphone should be placed on the right side of the harp, from the perspective of the player. One common mistake with recording harp is to place the microphone while the harp is in its resting position. Harpists tend to tune the instrument in this position, as it is easier to reach the tuning posts while standing up. As the harpist prepares to play, the instrument is tilted in toward the player, and the distance from the microphone to the soundboard may increase by as much as 30 cm or 1 ft. This will result in an unsatisfactory harp capture, along with excessive orchestra leakage in the harp pickup.

As with the percussion instruments and piano, the harps are frequently delivered and set on stage by the crew long before the musicians arrive for the recording or rehearsal. It should be the strict policy of the recording team that the instruments are not touched unless permission has been granted by the stage manager or the musicians themselves.

Figure 9.2 Microphone Placement for Harp in the Orchestra

EXERCISE 9.1 PLACING THE HARP MICROPHONE

1. Start by preparing the harp microphone on a stand with a boom, leaving it at a safe distance from the harp until the harpist is on stage. Additional space should be left if the harp still has its cover on to avoid damaging the microphone. Once the harpist is ready, the microphone can be brought in and placed while the harp is held in playing position. This is quite different from the resting position, which results in a big difference in microphone placement.
2. Check that the harpist has enough room to comfortably restore the instrument to its resting position when not being played. Great care should be exercised to ensure that the microphone stand does not tip over and fall into the harp.

3. A sandbag can be placed on the base of the stand, especially if a large, heavy microphone is in use. Otherwise, it is essential to ensure the stand and its boom arm are properly tightened and balanced to avoid damaging the harp. This can be helped by placing the base of the stand as close as possible to the base of the harp, then using less extension of the boom to realize the desired microphone position.

9.2 Orchestra Piano

This section deals with the piano as an instrument of the orchestra, as opposed to its role as a solo instrument – see Section 11.4 for detailed information on recording piano concertos. Orchestral piano, as one instrument in a large ensemble, is generally meant to blend rather than dominate. Placement can therefore be near the rear of the stage, behind the violins, or somewhere on the left or right side of the woodwinds. If possible, the main body of the piano sound should be present in the main pickup, so that spot microphones are only needed for detail and for minor balance corrections.

Preferably the piano should be positioned so that the lid can be fully open, but usually this is a problem for other musicians' sightlines to the conductor. It may be better for recording to completely remove the lid rather than using the tiny "half stick", which is actually much shorter than half the length of the "full" stick. Once the lid is off, very little sound will be projected forward, so the microphones should be placed up and over the soundboard. Even at 60 cm or several ft. above the strings, the piano will be incredibly direct sounding, so issues with leakage will be minimal.

EXERCISE 9.2 PLACING THE ORCHESTRAL PIANO SPOTS

1. One microphone positioned in the middle of the open instrument may be enough to capture detail, although two microphones will offer greater coverage of the instrument's entire range. The use of two stands rather than a stereo bar on one stand may offer greater flexibility in positioning and control; however, one stand with a bar will take up less space on stage.
2. Placement should be slightly offset from the hammers and just outside of the instrument in order to achieve a capture that will blend well with the main system and doesn't sound overly present or close in the overall balance.
3. In the mix, care should be taken to portray the instrument so that it is actually recessed into the orchestral soundstage. This way it will not be misrepresented as a solo instrument in a concerto setting.

9.3 Celeste

While it might seem like an interesting idea for the recording to separate the celeste and piano, or to place them on opposite sides of the stage, the recording team should first check whether there are two separate players or one player covering both parts. In this second scenario, the instruments will need to be placed next to each other. The warm tone and moderate volume of most celeste instruments allows for a fairly tight placement of the microphone, although the mechanical knocking sound of the keyboard will be quite obvious at very close proximity. The best sound generally comes directly out of the back of the instrument. Placing the microphone at a distance of around 60 cm or 2 ft. from the back of the instrument should result in a capture with adequate control and with a manageable amount of leakage from other sources. A certain minimum amount of leakage or bleed is necessary to allow for a natural blend and live-sounding presentation in the mix.

9.4 Organ

Balancing organ and orchestra is very much an acoustic process that needs to be addressed within the hall. There isn't much to discuss here, as most halls with an organ will require that the balance be controlled by the conductor and the organist, as the organ, when played loudly, will be in all microphones in the room, and impossible to control over the microphones. Unlike the occasional harp glissando, an organ passage that risks drowning out the string section or the entire orchestra is less desirable, no matter how impressive the sonic experience might be. Certainly, the conductor can be offered guidance from the recording team in adjusting the organ balance from the podium, and through the playback of a soundcheck some volume correction can take place. Quieter passages needing support or focus can be helped by placing a pair of microphones closer to the organ.

With any luck, microphones to capture the organ will be permanently suspended in the hall in an adequate location, since that position is probably very high in the air. As there is little that can be done in a subtle and controlled manner, any rough microphone placement that is closer to the organ than the orchestra will help. It may be hard to reach a workable height even with very tall microphone stands, and temporarily hanging the microphones may be time consuming or practically impossible. A pair of microphones in the organ loft might help to add some clarity, even if the placement is rather close to the pipes, resulting in an uneven pickup.

9.5 Chapter Summary

- Be prepared to find these instruments in any stage position, as there is little convention for their placement.
- Capturing each of these instruments too closely will run the risk of sounding as if they were recorded separately.
- Wait to place microphones until after the piano lid has been raised, and the cover is off the harp, so as to remain out of the way of the stage crew and musicians.
- The organ will not be easy to adjust much beyond the natural acoustic balance in the room.

Part III

Variations on a Theme

Chapter 10

Recording Chorus

This chapter groups together all scenarios where a chorus is concerned. All the relevant information and discussion is presented in one place, because of the many overlaps in methods for recording the various iterations of choral ensembles. Solutions are presented for recording chorus on its own, chorus with piano and chorus with orchestra.

10.1 Recording Chorus A Cappella or Unaccompanied

Hearing a choral group singing in a reverberant space can be a memorable event, as multiple voices join together to create an immersive aural experience. Even a novice choral recording that is unbalanced and lacking in clarity can sound very beautiful, while a very well-recorded chorus can result in a thrilling, lifelike presentation. A cappella choral recording presents several difficulties that relate directly to recording orchestra and chorus together. Capturing each section in a manner that affords good control over the balance is certainly a priority. Having achieved that, it is the overall sonic picture of the chorus that is most important –presenting the ensemble so that no individual voice is "sticking out" in the overall mix. Smaller choruses are easier to record with clarity, as the diction and rhythmic ensemble in larger groups tends to become somewhat spread, or veiled, as a greater number of singers work to synchronize their collective performance.

Choirs are typically configured in four parts positioned left to right, separated into soprano, alto, tenor and bass voices. A children's chorus may be in two to four parts, comprised of only soprano and alto voices (or two soprano and two alto sections). In addition to a main system, one support microphone for each section is common (see Figure 10.1), unless the group is small (12–20 voices) and naturally well balanced, when a main system might suffice. More complex arrangements will call for split sections, and the recording team may find that the "firsts" are in front of the "seconds" in each section (e.g., Tenor I in front of Tenor II). With any luck these internal balances will be kept in check by the conductor, although the recording team should be ready to point out or adjust for any consistent issues or general tendencies.

Some choral directors are confident enough with their ensemble's talent that they choose to intersperse the voices so that all four parts are blended across the ensemble. For this configuration to be successful, the singers need to be very secure in their parts, as each neighboring voice will be singing a different line. Excellent results in intonation can be achieved in this way, at the risk of some shaky rhythmic ensemble and the occasional false entry. In this configuration the balance of the four musical parts will be up to the conductor and the chorus members themselves, as the recording engineer will not be able to separate the elements.

DOI: 10.4324/9781003319429-13

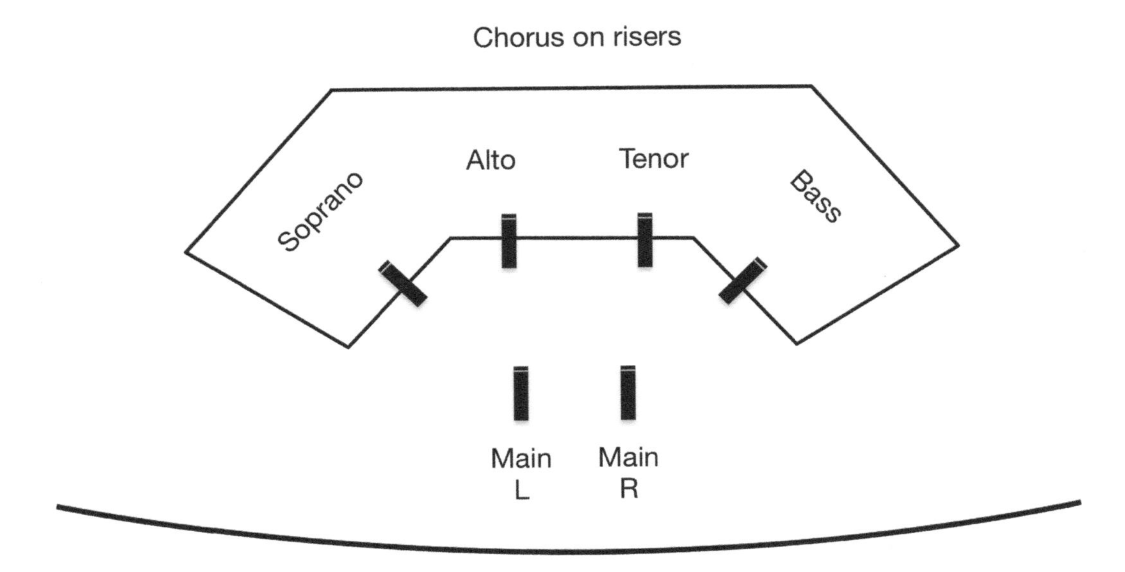

Figure 10.1 Diagram of Chorus on Risers, Showing Main Pair and Support Microphones

Risers for chorus tend to be terraced like gymnasium bleachers or a wide staircase, to allow all chorus members to be seen and heard by the conductor and the audience. A suitable capture will need to be set at a decent height, otherwise the front row of singers will be much closer to the microphones than the other rows (see Figure 10.2). Placing the microphones slightly "up and over" the group will help to reduce the front-to-back depth.

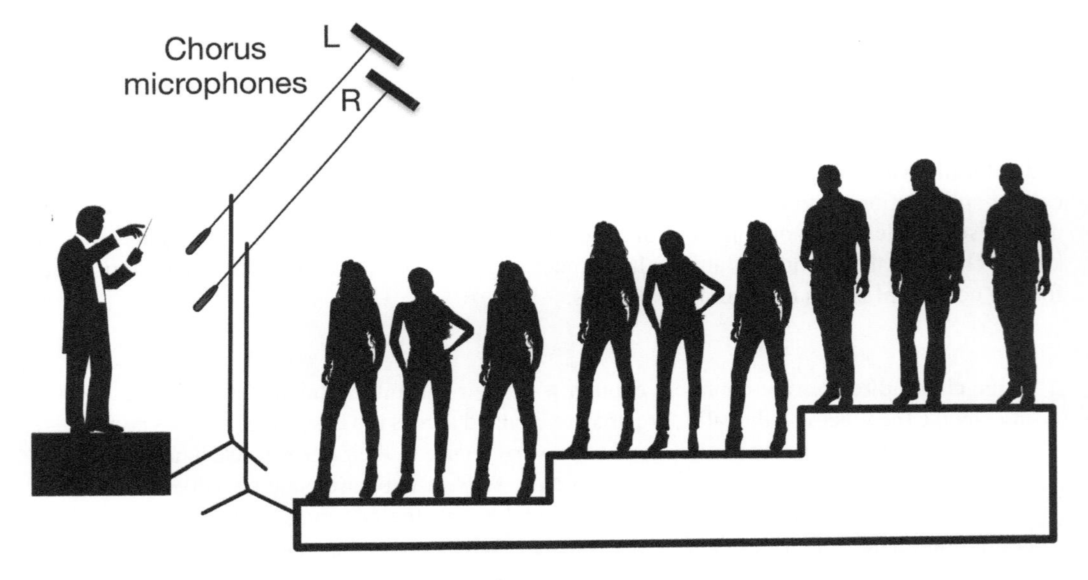

Figure 10.2 Diagram of Chorus on Risers (Side View) and Typical Microphone Placement

Directional microphones can be oriented so that the farthest and most upper row is on axis. At the same time, the level and presence of closer performers will be attenuated. The disadvantage is that the front row may be projecting slightly under the microphones, so that some of the body and core of their sound may be missed. Additionally, a brighter capture will yield a result that actually favors presence at the expense of fullness. Careful adjustment of the microphone placement during listening can help to even out the overall pickup of both the main microphones and the closer spot microphones on each section.

Once an ideal microphone placement is realized for the overall presentation, there may still be certain voices that stick out – even in the main microphone system. Listening in the room with the choir can be helpful in discovering where the issues lie. If there is an opportunity, the choral director might be asked to instruct a "stronger" member of a section to hold back so that they blend more naturally with the section.

Normally the director will know their choral group very well and will know which individual singers are the usual suspects when it comes to uneven section blend. Choral conductors tend to be experts in adjusting the volumes and positions of individual voices within the ensemble.

10.2 Recording Chorus with Piano Accompaniment

Recording chorus and piano presents its own unique challenges. Compared to the choral sound, the piano tends to sound both loud and distant or overly diffuse in the room, especially when the chorus numbers are significant (30 members or more). Moreover, it is quite common to find the lid of the piano set on the short stick, in a failed attempt at taming the stage volume of the piano (see Section 20.2) and to optimize sightlines between the conductor and the chorus members. Stage setups with the piano placed all the way to one side make for a highly uneven recording. For live concert recordings of chorus and piano, there isn't much that can be done, but it is a good idea to suggest that the piano be placed close to the center, if not center stage, in front of the group. If the use of the short stick is unavoidable, at least a pair of bright piano spots can be placed where some detail and clarity can be added. In the event that the piano is positioned well off to one side (normally to the conductor's left), there may a possibility for the lid to be fully open on the tall stick. There is an opportunity in this setting to use omnidirectional microphones on the piano, fairly close to the instrument, providing clear and full-frequency coverage of the instrument. Here the piano lid acts as a baffle, protecting the piano microphones from leakage coming from the voices. This is similar to the approach for recording piano concertos as described in Section 11.4. If the sopranos and altos are positioned on the left side of the group near the piano, those chorus microphones can be filtered to reduce the amount of low-frequency leakage coming from the piano.

In a recording session, it may be possible to place the piano behind the conductor facing the ensemble (see Figure 10.3). The main microphones would then be positioned between the piano and the chorus, so that a good amount of direct piano sound is present in the main system. The conductor may find this setup unfamiliar and difficult for communication, so they should be asked about considering this configuration well before the recording session takes place. In this setup, the

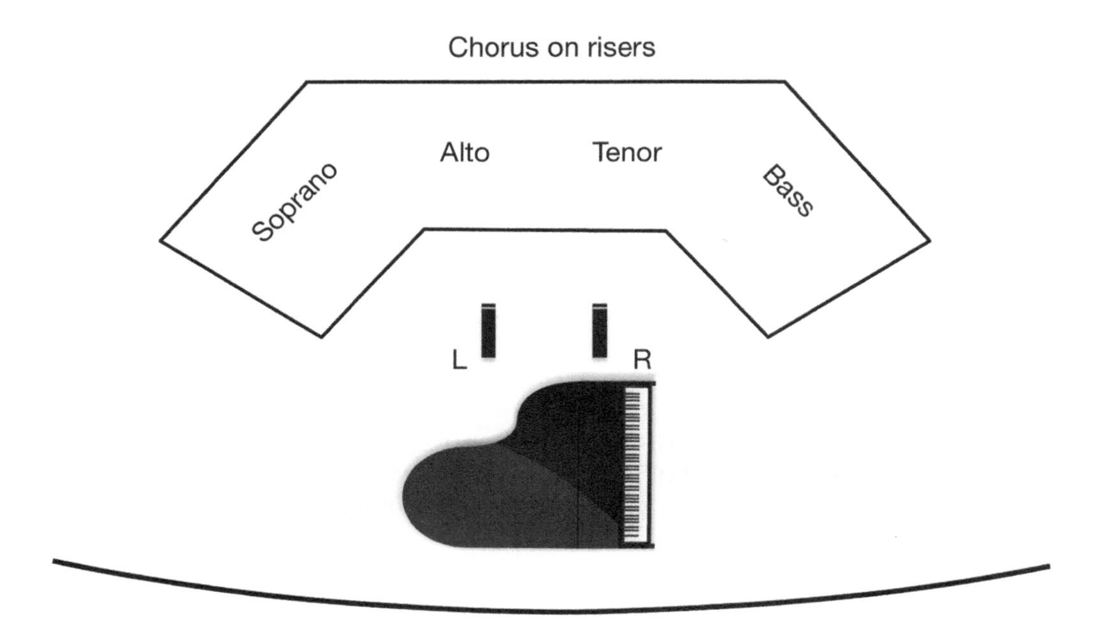

Figure 10.3 Diagram of Piano Facing the Chorus in a Recording Session Setup

mains should capture the fullness of the piano, so cardioid piano spots would be the best choice for adding clarity and stability to the image.

10.3 Recording Chorus with Orchestra

One of the most difficult orchestral elements to capture is a large chorus positioned at the back of the stage, either on risers or in a choir loft above the orchestra at around 12 m or 40 ft. from the conductor and the main microphone system. The challenge is to get enough presence of sound and diction (the words), while maintaining the blend of each section of voices. Also, the perspective of the chorus should remain stable and set back behind the orchestra, if that is the actual stage placement of the group. Leakage of sound from various percussion and brass instruments into the chorus microphones is a regular problem.

It is difficult to add clarity and volume to the chorus without "bringing them forward" as the chorus microphones are opened up in the mix. A normal perspective would place the singers behind the orchestra, although if this is done with any sense of natural presentation in mind, the chorus sound usually remains unclear and underbalanced. A few options can be attempted and assessed for their potential as solutions to this common problem, as suggested below.

Careful fader settings: the assumption here is that the chorus can be heard well in the main system, albeit with a general lack of presence. Fader gain is a very easy way to control perspective – push the level of the chorus channels until the point at which they begin to sound too close, then remain below that level in the mix. With any luck, this level will be adequate to present the chorus in a balanced way, and with good clarity. Care must be taken to manage the levels when adjacent instruments are playing, such as percussion or brass. The resulting leakage into the chorus microphones risks overemphasis of those sections in the balance.

Employing reverberation to the chorus channels in order to push them back: a good amount of artificial reverb on the chorus channels can help keep the choir from sounding too close, but this can be at the expense of diction and clarity if it is overused. Leakage of adjacent instruments will also be affected by the additional reverb and must be kept in check.

Using bidirectional microphones for more control: while cardioid polar patterns will offer better section coverage and blend, figure 8s can offer greater rejection of adjacent sound sources – particularly timpani and percussion, which tend to have loud passages when orchestra and chorus are in the climactic sections of a piece. The microphone capsules need to be positioned so that the null point faces unwanted instruments, which means the microphones will need to be somewhat up and over the chorus. This is not an ideal position for the choral sound, but it can be a workable compromise in certain repertoire where isolation is critical. As mentioned with woodwinds in Section 8.1, the figure 8's narrow polar pattern will make blend and section coverage an issue, so that higher microphone placements may need to be considered. A low ceiling over the chorus might also be an issue in this case, as the back of the figure 8 diaphragm will be susceptible to capturing unwanted ceiling reflections. It may be determined, however, that a good compromise might be to stay at a lower height for more control and separation of the chorus in the mix, rather than evenly capturing all the voices of a section. This will work well as long as there is a good amount of choral sound in the main system. The same can be said here for using cardioid polar patterns in a tighter-than-usual placement across the front of the chorus.

Placing the chorus in the audience seating area, out in front of the orchestra: this works well for recording sessions and is a great technique for sound and balance. The overall amount of chorus in the main system can be adjusted by the placement of the chorus (which rows of seats to use in the house), and the chorus microphones end up far away from noisy brass and percussion. If the chorus is much closer to the main system, less support will be needed from the spot microphones. The downside to this method is the difficulty of communication with the conductor. Even placing a camera on the conductor and video monitors for the chorus, the tendency will be for the chorus to be late, or not together in their performance unless they are very well rehearsed (and have rehearsed in this configuration). Obviously, this method would need to be proposed well before the sessions if it were to be attempted.

Delay compensation on the chorus channels: by offsetting the chorus inputs in time, their perspective can be kept the same as it appears in the main microphones. This is accomplished by delaying the chorus channels so that the time of arrival is more or less synchronized with the vocal sound as it arrives naturally at the main capture system.

For delay compensation of spot microphones, the calculation is around 3.3 ms/m, or 1 ms/foot, as sound travels at around 340 m/sec, depending on a few variables such as temperature and elevation. If the distance from the chorus capture to the main capture is 10 m or 33 ft., a delay of 33 ms can be applied to the chorus so that the signals are more or less time-aligned and the natural stage position of the chorus as presented in the main system remains intact.

EXERCISE 10.1 APPLYING DELAY COMPENSATION TO CHORUS/SPOT MICROPHONES

This exercise can be performed with any support microphones across all sections of the orchestra, including soloist microphones:

1. After all positions are set, measure the distance from the chorus microphones to the main system. When using three or more microphones, a different delay may be applied to each microphone for greater accuracy. The easiest way to measure the distance is to start recording, then excite the room by clapping once (or using some other impulse generator) in front of each chorus mic, from left to right. If more microphones are being measured, a simple voice "slate" can be added at each position. For example, say "English horn" before clapping in front of that microphone.
2. The resulting recording can then be used to measure the time difference between the close capture points and the main system. Any Digital Audio Workstation (DAW) can be set to read its timeline window in milliseconds. By clicking on the first impulse and dragging to the same impulse recorded on the main microphones, a time difference in milliseconds can be measured and set for each channel.
3. Apply the respective delay amounts to each close microphone channel and group the delays within each section so they can be bypassed together with a single keystroke or mouse click (for instance, grouping all chorus or all percussion channels together).
4. Compare the effect by listening to a section of the performance, toggling the delays in and out. Adjust the delays by ear for a more effective result, trying slightly longer or shorter delay times, respectively.

Leakage from adjacent instruments may be positively or negatively affected by the delay and should be evaluated accordingly. Also, according to the precedence effect, if the chorus microphone signals arrive slightly after the chorus is heard in the main system, adding delay may actually work against the effort to increase clarity [1]. A more accurate measure might be to use samples rather than milliseconds. This may be helpful when working with more critical issues such as soloist microphones at the front of the stage, where the distances between microphone systems are much shorter.

10.4 Chapter Summary

- Recording chorus alone has similar issues to recording chorus and orchestra together.
- The goal is to achieve a clear chorus sound without too close a perspective.
- Quite often the best way to solve internal chorus balances is to speak with the choral director.
- Placing the chorus in the audience seats can be a great solution, but with its own limitations.

Reference

1. Everest, F.A., and Pohlmann, K.C. (2009). *Master Handbook of Acoustics*, 5th ed. New York: McGraw-Hill, pp. 60–61.

Chapter 11

Recording Concertos

This chapter addresses placement of the soloist on stage and visual communication between the soloist and conductor, and also includes a general discussion of balance between soloist and orchestra. In this variation of orchestra recording, the conductor becomes the accompanist (though not all conductors would agree). In some cases, there may be political issues that arise during the recording of a concerto; specifically, differences of opinion between conductor and soloist regarding balance, tempo and overall musical interpretation. The recording team, and more specifically the producer, will need to be prepared to delicately navigate this landscape.

There is a famous story of pianist Glenn Gould and conductor Leonard Bernstein disagreeing on tempos in the Brahms D-minor Piano Concerto (No. 1). Bernstein actually told the audience before the concert that he didn't agree with Gould's choice of tempos, but clarified that he fully supported and respected the soloist's wishes, even though they were quite unorthodox [1].

While many references define the word concerto (or concert) as meaning *agreement* or *harmony*, Old Latin actually describes "concertare" as to *settle by argument*, or to *decide by fighting*, therein suggesting the concept of a musical struggle between the soloist and orchestra [2]. By this latter definition, it follows that there will be times in concerto performances where the orchestra is intended to overpower the soloist, even though the soloist is doing their best to be heard through often busy and dense orchestrations. In live concerts, the audience can take advantage of the visual element of the soloist on stage. When the listener watches the artist while they listen to the performance, their eyes help their ears discern the sound of the soloist from the texture of the accompanying orchestra, or at least trick the brain into thinking they are hearing the soloist at all times. In an audio recording of a concerto, the listener will require extra help to keep track of the soloist. Proper placement of the spot microphones will afford the engineer the ability to keep the soloist clearly out on front and well balanced at most times – except for those moments when the orchestra is expected to swallow up the sound of the lead voice. In earlier works such as the violin concertos of Vivaldi, the soloist is meant to blend into the orchestra sound during the tutti passages, temporarily adopting an equal voice as part of the string section.

DOI: 10.4324/9781003319429-14

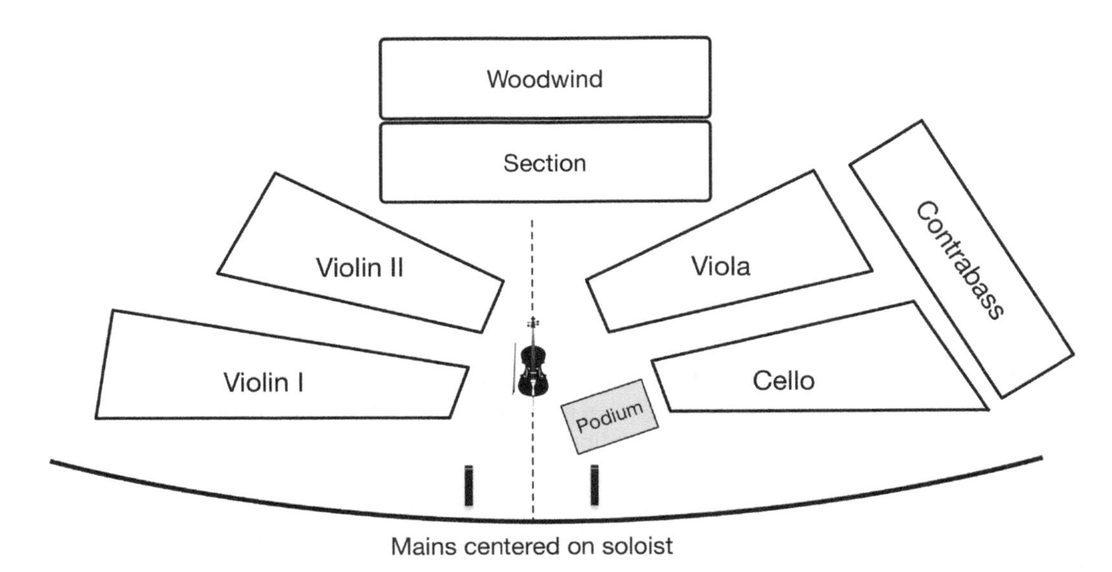

Figure 11.1 Diagram of Conductor's Podium Offset to the Right and the Mains Shifted to the Left in Order to "Center" the Soloist

11.1 Soloist Stage Placement

If possible, the soloist should be presented in the center of the recording, with the orchestra sound spread equally to the left and right. The most natural way to achieve this is to physically place the soloist "center stage". Some negotiation will need to take place in the preproduction phase, as the conductor and stage crew will need to agree to this configuration. For instruments other than piano, the conductor's podium should be shifted to the right so that the soloist may be positioned as close to the center as possible, as in Figure 11.1. In a live concert, space is made for the soloist by removing the stand of violins closest to the conductor and shifting the players in the section accordingly.

This may require adding one stand at the back of the section, but normally for a concerto the string count will be reduced from the regular number of players. Instead of a configuration that leaves a space for the soloist on the left of the conductor, a request can be made to move the podium to the right, along with the section of strings that sits directly to the right of the conductor (normally cellos or violas, sometimes second violins). The first violins may still need to be moved slightly to the left to create an opening that is large enough for the soloist to be comfortable.

It is not always possible to end up with the soloist exactly in the middle; it will depend on the size of the stage, or other constraints such as a large percussion setup that might be crowding the rear of the string sections on the right. In this case, the engineer might consider compromising the orchestral image by shifting the main system slightly to the left, so that the soloist is positioned exactly in between the left and right main microphones. Woodwinds, and perhaps other sections, may be slightly offset by this bias, although a small correction can be applied when bringing the woodwind spots into the mix at their "normal" positions. The artifacts of this correction are much less noticeable than trying to pull the soloist into the middle with their spot microphones, as the soloist's image in the main system will be very clear due to their close proximity to the front of the stage.

In a recording session, a soloist may prefer to be positioned at center stage inside the orchestra setup, directly under the baton of the conductor, as there is no need to worry about providing

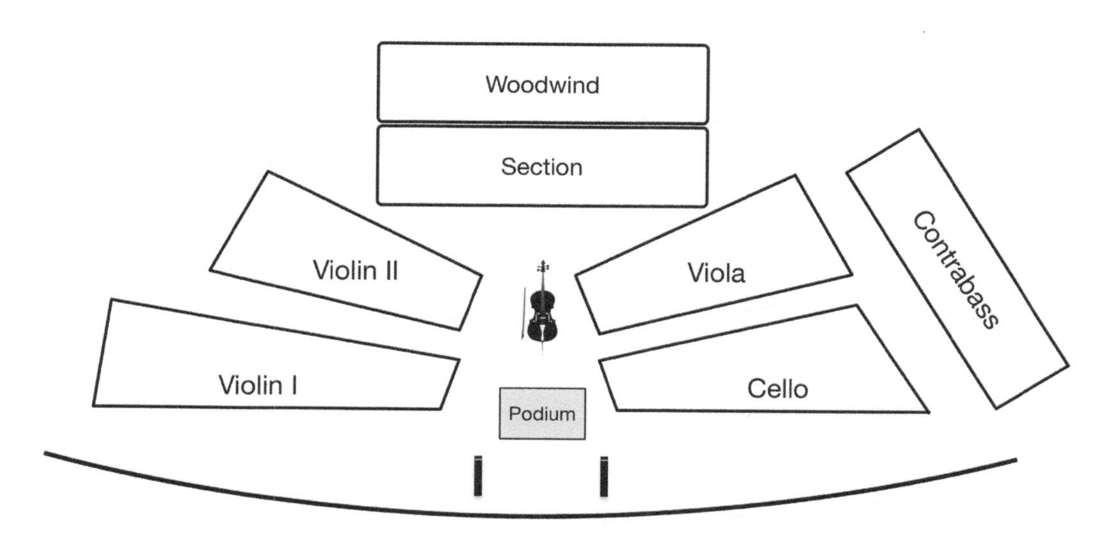

Figure 11.2 Diagram of Soloist Placement in Front of Conductor

decent sightlines for an audience (Figure 11.2). Alternatively, if there is room, the soloist may be placed out in front of the orchestra, looking back toward the strings. The conductor will invariably need to be rotated slightly so that they have adequate visual communication with both the soloist and the orchestra (Figure 11.3). This particular setup results in the rear of the solo spots facing the orchestra, yielding a greater amount of rejection of orchestra "leakage" into the soloist capture, and therefore more control over the balance. Soloists can generally hear themselves more easily in this configuration, which is an added benefit.

EXERCISE 11.1 SHIFTING THE PODIUM TO CENTER THE SOLOIST

1. Well before the recording session or first rehearsal and with prior approval from the conductor and stage crew, adjust the podium and string sections to make a space at center stage, large enough for the soloist to be comfortable.
2. When the soloist arrives on stage, ask if they are satisfied with the amount of space provided for them, including any nearby microphone stands and cables, the edge of the conductor's podium, etc.
3. Once they are comfortable, mark their position on the stage floor with tape (using a subtle color if it is a live concert recording) so that they will start off in the same place for each take, or each performance of the piece over multiple concerts.
4. The conductor will also need to be asked for their approval of the soloist placement (and resulting shifted string sections), as the soloist/conductor interaction is paramount to the success of the performance. The conductor needs to easily communicate visually with the soloist as well as the entire orchestra.

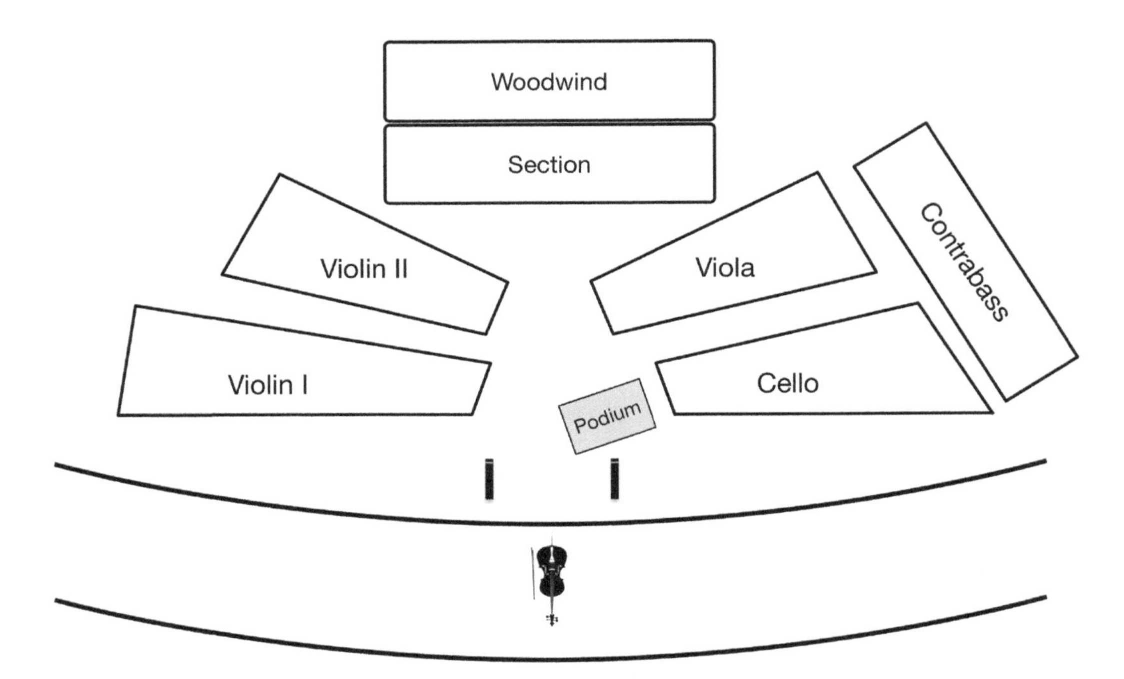

Figure 11.3 Diagram of Soloist Placement Outside the Ensemble, Facing the Orchestra

11.2 Main Microphone Placement

When positioning the main microphones and the soloist for a concerto recording, two issues should be considered: overall balance and orchestra image. Ideally for the balance, the soloist should be slightly low in the main pickup with respect to the orchestra so that a certain nominal amount of the spot microphones will always be needed. This will allow the spot microphones to provide a consistent perspective and stability of image. If the soloist dominates the main pickup, the orchestra risks being "dwarfed" by the soloist, and there will be no means to lower the soloist in level. As long as the conductor is closer to the audience than the soloist, as is generally the case based on their need for ease of communication, this should take care of itself. If the soloist is positioned out in front facing the orchestra, which is a common approach in a large studio or scoring stage, the main system should be placed between the soloist and the ensemble. In this setup, the soloist image in the main pickup can be slightly recessed by moving the artist's position farther away from the orchestra and the main microphones.

Always aim for the soloist to be slightly underbalanced in the main system, so that it requires support from the spot microphones. In the case of the "giant soloist" in the main pickup, the orchestra will sound small in comparison, and will need to be raised in level by using the various spot microphones throughout the ensemble. The net result will lack depth and overall weight in the orchestra sound.

The second consideration in positioning the main system is to decide upon the width of the orchestra image. When large forces are in play, or when the piece exhibits very dense or intricate textures in the accompaniment, it might be worth considering capturing the orchestra with a slight "hole in the middle" to leave room for the soloist. Simply placing the main microphones at a wider spacing will achieve the desired result – for example a normal spacing of around 1.5 m (5 ft.) might be changed to 1.8 m (6 ft.) in order to open up the middle for the soloist. For a concert recording where the microphone setup needs to accommodate multiple pieces (e.g., a symphony and concerto on the same program), either the microphones can be adjusted between pieces, or a center main microphone can be used to fill the center when needed, then reduced to a lower level in the mix (or not used at all) during the concerto piece.

11.3 Soloist Microphone Placement

Two identical microphones should be used to capture the soloist, even though it might seem as if one microphone would be sufficient to record a single instrument. Piano is of course the exception here, as it requires two channels anyway to properly capture such a large instrument. There are several reasons for using two microphones, and to do so with both capsules facing directly toward the instrument rather than using a technique such as ORTF or XY. This "narrow stereo" capture of the soloist will match the image of the resulting leakage of the orchestra with the main capture system. In practice, with the microphones positioned in parallel and spaced at around 12 cm or 5 in. apart, the resulting image of the solo instrument is virtually monophonic (very little difference in left/right amplitude or time of arrival), but the orchestra leakage into the solo microphones will be spread quite widely.

This pickup provides greater stability than a conventional stereo technique, as it can help to diminish the amount of a performer's left-to-right movement that may be exaggerated in the main system. For those moments when it is necessary to "dig out" the soloist, the orchestral image will remain stable – for example, the first violin section is not pulled to the middle as they would be in the case of a single mono spot panned to the center. The practice of placing both microphones on axis to the instrument offers an additional safety measure, in case one of the soloist's channels fails in the middle of a recording session or concert due to unforeseen technical problems. The recording can still be salvaged at this point by using the one remaining on-axis microphone. This would not be the case using an XY system, where losing one side would leave a single capsule at a 45° angle to the source.

In some extreme cases, one of the two solo microphones may be reduced in gain by 6 dB. This can be done as a safety measure for protection against overloads, at times when the soundcheck is too short or nonexistent, or when working with an unpredictably dynamic soloist, such as an opera singer. When auditioned as a stereo pair, the 6-dB difference in gain can be compensated for on the mixing console in order to achieve equal gain in the mix. When positioning the soloist microphones relative to the main microphone system, the distance between the two systems should be carefully assessed, as described in Section 6.1.

Standing soloists (violin concerto): for violinists, and that group of soloists who normally stand while playing such as trumpet and flute, marking three sides of a box on the stage floor will help to keep them in place, but also give them an indication of which direction they should be facing. All things being equal, if the violinist is facing the conductor, the instrument itself should be facing more or less directly into the hall and the main system. Other solo instruments may need to be asked to project more toward the hall and the main system rather than toward the conductor. Large stands with booms should be used, so that the soloist has plenty of room to perform comfortably. For live concerts, the solo microphones should be hung so as not to block the audience's view of

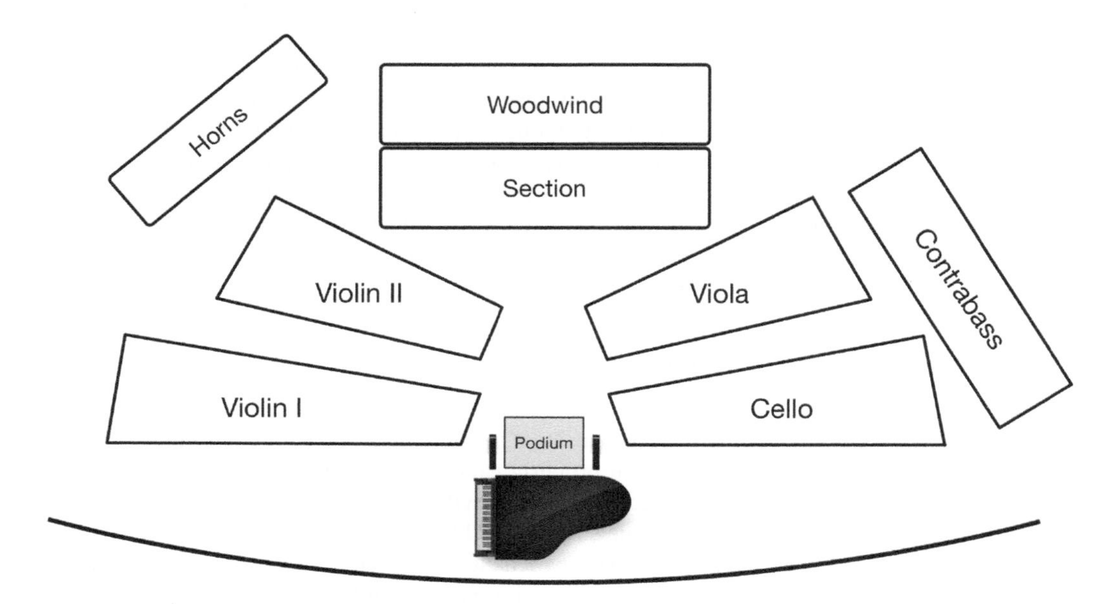

Figure 11.4 Diagram of Main Microphone Placement between the Piano and the Orchestra

the soloist. If hanging is not possible, then the microphones should be placed on a short stand and set to "angle up" toward the soloist.

Sitting soloists (cello concerto): as with harp, mandolin or any soloist that sits, it might be a distinct advantage to place solo cello on a small riser to raise the instrument up closer to the main system. Marking the end-pin position will get the cellist back in the same place before each take, and marking the chair position of a soloist such as mandolin will act as a decent guide for consistent imaging.

11.4 Recording Piano Concertos

The approach for a piano concerto will be quite different from the other setups. The conductor's podium will be placed farther upstage than usual to allow for the instrument to be positioned between the conductor's podium and the front edge of the stage. The piano keyboard will be placed more or less "center stage", so there will be no need for the process of offsetting the podium, as described for other soloists. In order to avoid a balance that is biased toward the piano and presents a recessed image of the orchestra, it is best to position the main system farther upstage between the piano and the conductor, or slightly over the conductor if necessary (see Figure 11.4). This placement will help to ensure that the orchestra is stronger than the piano in the main microphones, although in a very reverberant hall the piano may appear somewhat diffuse as compared with the orchestra. In this case, a substantial amount of piano spots will need to be present in the mix in order to create a suitably matched perspective.

For piano concertos it is recommended that the solo microphones be omnidirectional, so that a full and open-sounding capture is achieved, representing the piano in the most natural manner. The use of directional microphones is not required since the lid of the piano will help to reduce the amount of orchestra sound in the piano capture. Because the piano image will be very wide in the main system, it is a good idea to compensate for this by setting the piano microphones at a fairly narrow spacing, around 50 cm or 20 in. wide.

The width of the piano image should be presented in relation to that of the orchestra, so the piano should not be expected to fill the entire lateral soundstage. The resulting image will be narrower than what is normally suitable for a solo piano recording, as covered in Chapter 21. Also, the microphones should be placed quite close to the instrument, for greater control over the balance and to compensate for the mostly diffuse sound of the piano in the main system.

Piano concerto recordings in very live spaces are actually good candidates for two sets of main microphone systems (see Section 6.1), where the small "ab" might be placed upstage of the piano and above the conductor, and the big "AB" out in front of the piano for focus and clarity in the main system. Cardioid microphones as piano spots will probably work best when using this approach.

11.5 Chapter Summary

- Try to get the soloist centered on stage, even in a live concert setting.
- Be aware of artistic differences between conductor and soloist.
- Two microphones should be used to record a single solo instrument.
- The solo instrument should be slightly weak in the main system so the balance can be controlled and refined in the mixing console.
- Piano should be approached differently than other solo instruments.

References

1. Transcription of Bernstein's speech, Apr. 9th,1962, in New York, accessed July 12, 2023. https://web.archive.org/web/20001031125032/http://www.rci.rutgers.edu/~mwatts/glenn/lennie.html.
2. *American Heritage® Dictionary of the English Language, Fifth Edition.* S. v. "concert," accessed July 12, 2023. http://www.thefreedictionary.com/concert.

Chapter 12

Recording Solo Voice and Orchestra

As with any concerto recording, visual communication between vocal soloist and conductor is very important. Balance is even more important if we are to consider the fact that we now have text in a song that needs to be understood, where a violin melody could be slightly obscured for a few notes here and there with less issue. The basic mechanics and limitations of the voice must be considered, in terms of endurance, volume and power. This will affect the physical placement of the singer on stage, how long the singer or singers can perform each day, and the number of times a very demanding passage may be performed during a recording session.

It is important to bear in mind that vocalists are producing the sound of their instrument from within, and as such their performance will be "self-evaluated" on a more personal and possibly more critical level than might be the case for an instrumentalist. For instance, it is much easier for pianists to emotionally separate themselves from a mediocre performance by blaming the condition of the instrument – "*I normally play the Scherzo much faster and with more clarity, but the heavy action of this particular instrument is working against me*". Although temperature and humidity levels on stage (and backstage) are important considerations for any performing artist, they may present a greater concern for vocalists over instrumentalists.

12.1 Stage Placement

Much of the content in this section mirrors that of the previous chapter on solo instruments with orchestra. The desire to present the soloist in the center of the recording is still a priority, with the orchestral sound spread evenly from left to right. Leaving a hole in the middle of the image by setting the main system at a wider spacing than usual creates a "pocket" where the voice may reside. Less solo microphone level will then be needed to achieve good balance and clear diction with a reasonably natural perspective.

> In an effort to create the "illusion of reality", the vocal presentation should be somewhat larger than life. Care should be taken, however, not to lift the voice too much out in front of the orchestra in a quest for perfect diction – the soloist should remain part of the overall stage perspective, even with the extra support from the spot microphones.

The same preparations as for any other kind of soloist with orchestra should be employed, in that the orchestra and conductor are consulted ahead of time about offsetting the podium and adjusting

DOI: 10.4324/9781003319429-15

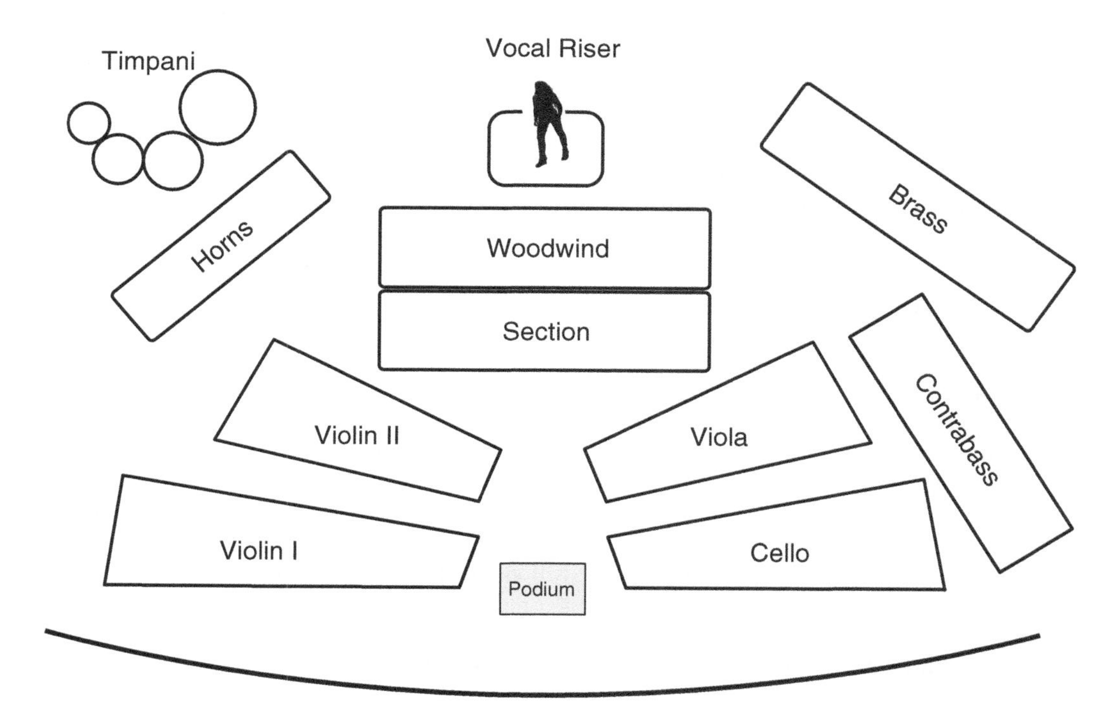

Figure 12.1 Diagram Showing Vocalist Placement at Back of Ensemble, at a Great Distance from the Main System

the stage plan so that the singer will be as close to center as possible. In a concert recording, the same stage change can be made for vocal solo with orchestra as for a concerto performance.

Placing a singer inside the orchestra setup may not be the preferred approach, as the soloist might have trouble hearing and the vocal leakage into all the string spots will be quite strong. Traditionally in recording sessions, singers have been placed at the back of the orchestra on a riser, simulating a vocalist on stage with orchestra in the pit, as is the common setting for opera performance (see Figure 12.1). While singers are mostly comfortable with this familiar configuration, there are several drawbacks. Because of the great distance from the conductor, singers tend to be rhythmically late or behind the beat; also, communication with the conductor between takes is difficult without the use of an intercom or telephone system, which in itself can be cumbersome and stressful to manage.

A major acoustical issue with this placement is that the vocal sound will be fairly diffuse in the main system, depending on the liveness of the recording space. The solution to this is to use more level from the vocal spots, at which point the soloist will risk sounding too close in the mix. Even with a substantial amount of close microphone signal in the balance, the "ghost" of the distant voice in the mains may present an obvious blurring effect, or at the very least, the impression that some strange processing was added. In terms of delaying the solo microphones for time coherence, the result may be a vocal performance that is too distant and of substandard quality due to the increased rhythmical "lag" caused by the delay.

During a recording session, having the singer face toward the ensemble on the outside of the orchestra and the conductor can provide good results. This allows the soloist to hear quite well and offers the orchestra better feedback from the vocal performance and therefore a better shot at a coordinated outcome. The other clear advantage is that there is less orchestra in the vocal microphones, which can be a lifesaver when working with less powerful voices. Having less voice in the orchestra spots can help with clarity as well, as a louder voice will manage to "pollute" all of the string section microphones, possibly causing an obvious comb-filtering effect when combined with the voice pickup.

12.2 Vocal Microphone Placement

As stated in the previous chapter, the best practice for capturing voice is to use two tightly spaced directional microphones, facing in parallel, for all of the same reasons put forth in Section 11.3. The signals are then equally applied to the left and right channels of the stereo mix, that is, panned "hard left and right". For live concert recording, there is even more of a case to reduce one microphone by 6 dB, as most singers exhibit a very wide dynamic range.

A stereo system such as XY or Blumlein (Section 3.4) in close proximity to a singer will most likely exaggerate rather than stabilize their movement during the performance – and if the right side of the pickup happens to fail during a live concert, the remaining microphone will be pointing toward the violin section rather than soloist.

Singers tend to move their heads while performing (and therefore their instrument) more than any other soloists. Briefly turning to look at the conductor and then back to the audience can cause a large image shift as well as timbral change as they move off mic and back again. A tight solo pickup introduced into the mix will help keep the image in the middle and stabilize timbre. When a vocalist wishes to orient their position so that they face the conductor for better communication, be very careful about offsetting the vocal microphones. If using bidirectional microphones, make sure to assess what signals might be captured by the backside of the capsules.

Warning: Capturing the soloist with a Blumlein system that is slightly rotated in orientation to the orchestra can create a major problem with the overall image – as the system is turned to the left, the back side of the left channel capsule will be exposed to the strings on the right side of the orchestra (presumably the cellos or violas) – and they will be copied over to the left side of the recording.

Reflections from the voice off the music stand can be a major problem. The comb filtering caused by the combination of the direct and reflected sound into the microphones will be more obvious on a vocal than an instrumental recording, especially on breath sounds. Great care should be taken to position the music stand and the microphones to avoid recording any sound reflecting off the stand (see Figure 12.2). A lower placement of the microphones can help to avoid this unpleasant artifact.

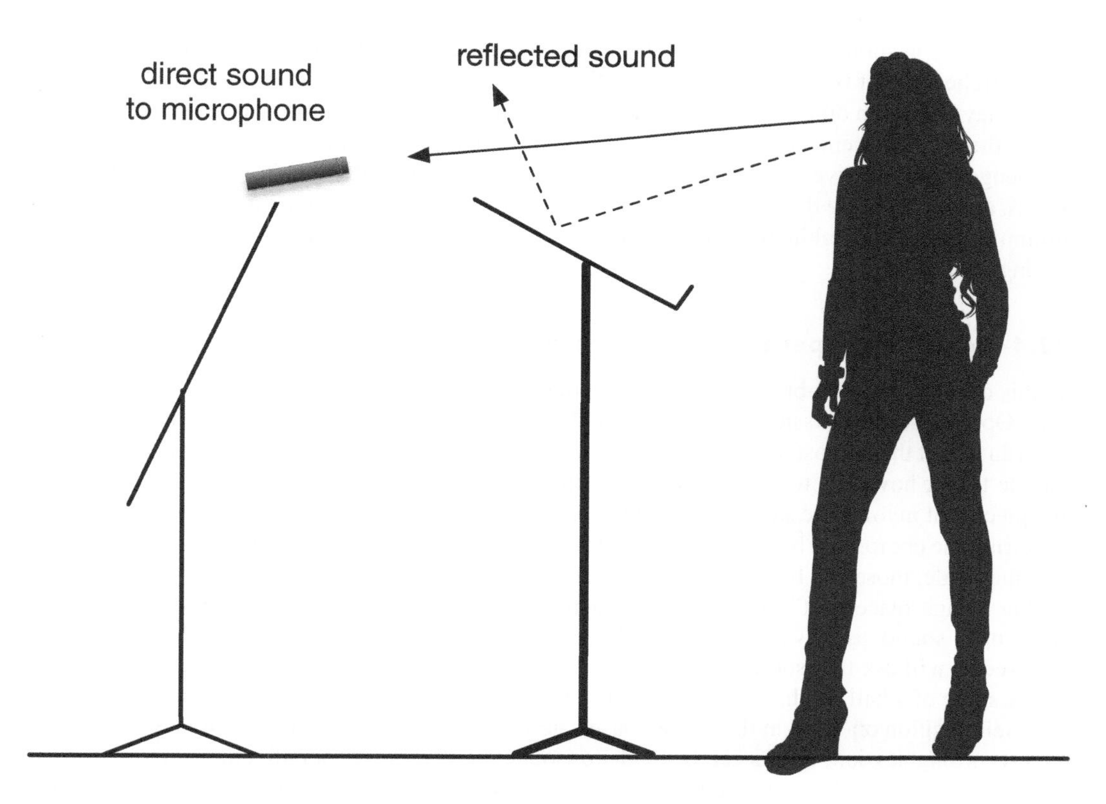

Figure 12.2 Diagram Showing Best Microphone Position for a Vocalist Using a Music Stand

EXERCISE 12.1 SETTING UP THE VOCAL SOLOIST POSITION

1. After the soloist is in place and has adjusted their music stand accordingly, check to make sure the angle of the music stand is such that it is directing any reflected energy away from the microphones. Always alter the microphone placement rather than the stand, as the singer will have set the stand to their own preference. Many recording studios cover their music stands with a layer of felt – this helps absorb some of the reflections, and it reduces noise from page turns and pencil dropping.
2. Check that the singer didn't adjust their stand after the microphones were placed; adjust again as necessary.
3. Keep in mind the discussion of dual systems (3:1 rule) in Section 6.1 when positioning the vocal microphones relative to the main microphone system.

12.3 Multiple Singers

Having more than one solo voice with orchestra increases the risk of problems with comb filtering, as each soloist is captured in the others' microphones and other adjacent spot microphones.

Live opera in a "concert performance" setting as opposed to a "concert staging" will present this problem, and in some cases, there may be six or more singers performing in a line across the front of the orchestra. For two voices, a good solution is to place the singers either side of the conductor. They may hear each other less easily, but the recorded sound will be improved, and they will both enjoy the close placement to the conductor. Repertoire with more than two voices creates more challenges, and in a live concert recording it is a matter of constantly monitoring the soloist fader levels, always favoring the microphone of the vocalist who is singing at the time. This effort should minimize the ugly combination of multiple phase relationships through careful gain management during the mix.

12.4 Recording Opera in Studio Sessions

In this day and age, full-blown opera productions recorded in the studio or an empty hall are very rare. Opera recording sessions can be a scheduling nightmare and a very expensive undertaking, to such an extent that almost all new opera recordings are live (see Chapter 15 for more details). Steps can be taken, however, to streamline the production, by organizing each session into sections of the piece that involve the same soloists and orchestra forces. For example, all sections of the work requiring the chorus can be scheduled on the same day, and if one soloist only sings in the opening and the finale, those can be scheduled in the same session.

For "stage placement" in a studio production of an opera, the orchestra will benefit from a more open sound, as they aren't crammed into the pit below the stage. Quite often the conductor or director will ask that some staging or risers be installed across the back of a recording studio, or the stage of a hall. In this way the singers can be set up behind the orchestra on risers, simulating their position on stage in the opera house, "above" the orchestra, as discussed in Section 12.1. In a recording session, efforts can be made to isolate the voices without affecting communication between the singers. Placing "gobos" or baffles between soloists will help to separate the microphone capture of each voice, but not the actual singers (see Figure 12.3). This will aid in reducing the amount of "crosstalk", or each performer's voice leaking into the microphone of an adjacent singer. When placing soloists across the back of the orchestra, gobos can also help to reduce the amount of reflected sound reaching the main microphones so that a cleaner overall sound might be achieved.

It is important that the soloists can still hear each other without amplification, as the practice of beginning an opera session by handing out headphones to the singers is not recommended. Unlike a musical theater (Broadway) cast, opera singers are used to hearing themselves and the orchestra acoustically on stage, so it is important that they continue to perform in this familiar manner in the studio. That being said, it is quite common that someone's part or some element will need to be overdubbed during the course of the sessions, as schedules change, or one singer may become ill and not be in "good voice" for a particularly strenuous passage. The recording team should be ready for this and have a headphone cue system ready, since the easiest way to match the sound of the overdubbed elements is to record them in the same room as the original material. Careful notes and measurements should be taken regarding musician and microphone positions, preamp gains, etc., in case it is necessary to come back at a later time when a singer is more healthy and ready to perform their high notes.

Certain sound effects or unique instruments may need to be recorded separately as well, for ease of control and options in balancing the mix. For instance, the actual anvils in the "Anvil Chorus" in Act II of Verdi's *Il Trovatore* are frequently overdubbed, so that their intensive sound doesn't obliterate the orchestra and distract the chorus from performing well and with good intonation.

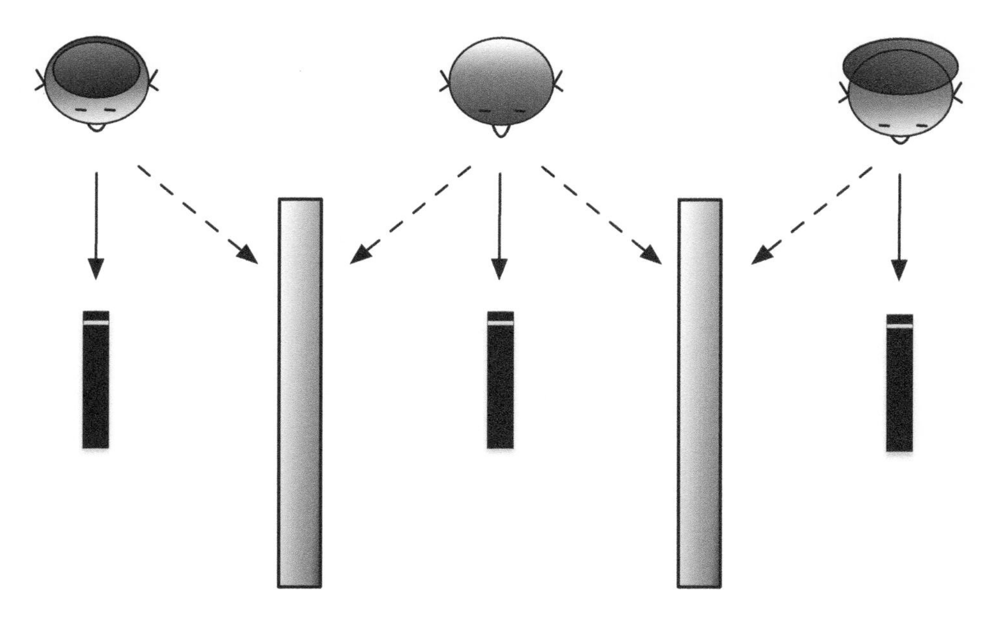

Figure 12.3 Diagram of Several Singers in an Opera Session with Gobos Separating the Spot Microphones

12.5 Creature Comforts

Voices tend to be more sensitive to temperature change than instruments. While sudden changes may affect the tuning of an instrument, even minor fluctuations in temperature and humidity can affect a singer's performance. For vocal recordings it is important to check the temperature in the hall or studio as well as the artist lounge or dressing room. This should be done well before the artist arrives so that there is ample time for adjustment, as air conditioning is commonly considered a "voice-killer". Most singers will bring water or tea with them, but it is a good idea to ask ahead of time if they need anything provided, such as a kettle or a small humidifier. Room temperature water is usually preferred, so it is recommended to keep the water out of the refrigerator.

12.6 Chapter Summary

- Try to get a single vocalist centered on stage, even in a live concert setting.
- Singers tend to be more personally connected to their "instrument" than other soloists.
- Be sure to check the temperature on stage and in the dressing rooms, making any adjustments well before the singer arrives.
- *"Life is short – Opera is long."* (Original source unknown; earliest use may have been a caption on a T-shirt in the 1980s.)

Part IV

The Recording Session

Chapter 13

The Role of the Producer in Classical Music Recording

Very few people outside of classical music recording actually understand the role of "producer", and their responsibilities during preproduction, recording and the postproduction stages of editing and mixing. For smaller-scale projects and recordings with limited budgets, it is becoming more and more common for one person to manage the roles of both producer and engineer. In the world of popular music recording, the producer's job description is generally very different. In this chapter, the classical producer's role is fully explained, and general advice is offered for those engineers and musicians who wish to develop a career in classical music production. The relationships between both engineer and producer and the producer and artist are addressed in detail.

First and foremost, the producer is normally responsible for overseeing all stages of the recording process and delivering a finished master of the final product to the client (either a record label or the artist themselves). In almost all cases, the producer is the lead member of the recording team. While the engineer might commonly be hired or requested by the producer, a label or artist might approach the team members separately in terms of availability and budget concerns. In either case, the engineer should be ready to consult the producer on most decisions involving scheduling and rental costs, as well as any major technical considerations, because in the end the producer is ultimately responsible for the outcome.

13.1 Qualifications of a Classical Music Producer

A well-qualified candidate for the field of classical music production is someone with a formal education in music and many years of performance experience. A musicologist may know the repertoire, but if they lack playing experience they will be unable to closely relate to the artist's life on stage or the hours in the practice room, as well as many of the physical considerations of a busy performer.

Understanding recorded sound is the second most important requirement. Some years of experience with critical listening and analysis is crucial when it comes to forming an opinion of a musical performance captured by microphones and reproduced over loudspeakers or headphones. This is important when it comes to evaluating the overall presentation of an audio recording, and for monitoring consistency in the sound throughout the process of a recording project.

The record production process must be fully understood as well, specifically the advantages and limitations of the editing and mixing phases, and how these may affect decisions made during the recording session. A very simple example is that when editing between two "takes" or versions of a particular passage, the tempos will need to be the same – otherwise a noticeable shift will occur at the desired edit point in the music. The producer can reduce this artifact by carefully monitoring the tempo of each take during the recording session. Certain irregularities in balance might

DOI: 10.4324/9781003319429-17

be adjusted during the mixing, while drastic differences in dynamics and intensity between takes might cause problems in the editing and mixing – the producer will need to be aware of this and take appropriate notes during the recording.

The producer of a classical music recording may also be tasked with creating a financial budget, including the cost of the hall, the musicians, the engineer, equipment and travel, and an estimate of the hours or days needed to complete the postproduction (editing, mixing). These additional tasks are now asked of the producer more often than they were in the past, as record labels are shrinking, and more artists are financing and managing their own recording careers. In certain cases, the producer may be involved in preparing contracts for the musicians, but only where the appropriate expertise exists.

Almost all professional orchestras are unionized, and as such there are strict rules governing recording session proceedings. Even nonunion ensembles will have a preferred structure as to how they record, and the producer needs to be informed. In North America there are strict rules in place for recording orchestras that have bargaining agreements with the American Federation of Musicians (AFM). Many organizations have negotiated individual agreements, so it is important that the producer consult with the orchestra management concerning the specific rules. In Europe the rules are different again, but in general there will be set record times and break times, as well as a limit on how long the orchestra can be asked to play in one sitting.

Additionally, there are limits on how much recorded material can be used from each session. This should be carefully assessed, since it will determine the minimum number of sessions that need to be booked for each project, thereby affecting the overall budget. Tuning is normally performed on record time rather than break time, except for the first tuning of each session, which is done just before the session begins.

The clock is always running, whether it be tracking record time or break time – the dual clock used to track time during competitive chess matches comes to mind. A certain amount of skill is required of the producer to manage session time, as precious recording time should not be wasted on playbacks or lengthy discussions.

Running playback sessions during the breaks is an efficient use of the time, since the conductor will need to hear at least some of what is being captured. This will require the producer to choreograph the timing of breaks so that playbacks happen at the right times. For instance, after series of complete takes and small inserts have been recorded to finish a piece or movement, it makes sense to record a complete run-through of the next movement, rather than calling a break. This way, the conductor will have new material to listen to during the break, and a few minutes for any discussion with the producer.

13.2 Relationship Between Producer and Engineer

In the most basic sense, the division of roles is that the producer is in charge of the music and the engineer is in charge of the sound – but in reality, there is a great deal of overlap. In most cases, this overlap will be biased toward the producer being actively involved in the sound, more than the engineer offering their opinion on musical issues. Engineers with a strong musical background will need to accept this inequity as part of the job description and be ready to collaborate with the producer (and the artist, for that matter) in creating the best sound for each project.

A team effort yields the best results, according to the old adage "two heads are better than one"– or, in this case, "four ears are better than two". A recording team that is focused on working together should be capable of realizing superior results on a more consistent basis than any individual endeavor.

The producer and engineer should maintain a "clientele" relationship. It is important that the engineer address all the producer's concerns, since the producer is normally responsible for choosing the engineer in the first place. The engineer will need to be ready to collaborate with the producer regarding microphone choices and placement, as well as the usual balance requests. Some producers will want to be more involved than others when it comes to decisions that affect the sound. By contrast, there are those who will leave technical issues, such as overall timbre, perspective and balance, entirely up to the engineer. This might seem very liberating – except that in the case of an unhappy artist, that same producer might not be any help in buffering the artist's frustrations, or offering any suggestions on improving the sound, and so the engineer is left to fend for themselves.

A more even division of roles will develop over the course of several projects, and as the engineer earns the trust of the producer, musical issues may begin to be discussed more openly. For example, the producer might ask the engineer which performance of a particular passage they prefer. It is good practice for the engineer to wait until they are asked for their opinion when it comes to musical considerations. Truly, the best one-line advice for young aspiring audio engineers is: "know your place". More importantly, the engineer should never offer up a musical opinion in the presence of the artist, and certainly not one that conflicts with that of the producer. This goes for issues of sound as well – best to wait until the artist has left the control before discussing any issues, and then present a united front at an appropriate time. A recording team that is seen to be always "in sync" will help bolster the artist's confidence and improve the outcome of the project.

There is a famous joke concerning music production, although certain members of the recording team may not appreciate the humor: *"Every recording is either well produced or poorly engineered"*. In other words, the producer receives all the credit while the engineer is obliged to accept all the blame, based on the traditional hierarchy of the roles.

13.3 Relationship Between Artist and Producer

While production styles may vary greatly between individuals, the most rudimentary function of the recording producer is to recognize and support the artist's vision for a particular performance, and subsequently guide them through the recording process. The description of "artist" extends to the conductor of an orchestra, a soloist or a group of musicians such as a string quartet. Guiding the musicians can be a very delicate process, and the producer needs to be well aware of their personalities and emotional states. It can be difficult for anyone with a strong musical background to suppress their own musical taste, but this may become necessary during the course of a recording session.

The producer should be very careful not to try and transform an artist's vision. There may be times when there is disagreement over performance practices, and the producer may point this out, but any proposed changes affecting musical style should be presented to the artist as suggestions or questions rather than explicit directions. It is important that the final interpretation of the recorded work is that of the artist, even though the producer contributes significantly in terms of decisions made during the editing and mixing process. One simply has to get "on board" with the artist and help them present their best version of the piece. After all, an artist that feels "bullied" rather than "guided" through the recording process might decide to look for a different producer on their next project.

The artist's state of mind needs to be constantly evaluated during the recording sessions. As such, a highly diplomatic style of communication will help keep the artist on track, as a great deal of soul-searching and self-evaluation will be taking place. The producer needs to help keep the artist focused on the task at hand, rather than dwelling on any negative aspects of the process.

One underrated skill all producers need to master is "talkback etiquette". This goes beyond the task of simply being sure that the talkback microphone switch is depressed before speaking to the musicians on stage, and that the switch is then released so that any private control room discussions are not "broadcast" into the hall. The all-important function of maintaining a constant dialog with the artist while they are out on stage or in the studio will help to ensure that the artist stays focused and feels supported by those in the control room. Certainly, the communications will need to be concise under the time constraints of a union orchestral session, but the producer should at least say something positive after each take. This is even more important when recording smaller groups or solo artists. A musician who is sitting in an empty hall after performing a complete take of a piece may end up feeling completely unsure of themselves unless some encouraging words from the producer are heard after the final chords have subsided.

13.4 Preproduction and the Role of the Producer

In order to be properly prepared for a recording project, the producer will need to be as familiar with the music as possible. Studying several existing recordings of the same piece is vital for the producer to have a sense of the range of available interpretations. This involves tempo, dynamics and balances between instruments or sections of the orchestra. In the case of a previously unreleased work or new composition, the producer should ask for a rehearsal recording or recent broadcast of a live performance, if one exists. At the very least, a computer-generated demo of the work might be used to help with learning the music. Knowing the rough timings of each movement or each piece can help in planning the recording session times.

Studying a piece ahead of time can help the producer (and engineer) plan how they might want to position the ensemble for the best balance and overall presentation. This may include deciding where certain instruments are placed under the microphones, so that they appear appropriately across the recorded soundstage.

Obtaining a copy of the score early on will ensure there is enough time to add markings and to study the score while listening to other recordings of the work. Highlighting features such as entrances, multiple systems on a single page, solo instrument lines and changes in clef in the middle of phrases can help with navigation of the score. Simultaneously following the score while taking copious notes is the epitome of multitasking, although this skill can be eventually mastered with enough practice. The more familiar the producer is with the written page, the more attention can be paid to the actual music. While wrong notes and misplaced phrases are usually obvious artifacts, missing notes in a piano chord or orchestral parts that are not played are less apparent in the context of a recording session. The producer cannot always rely on the conductor or the individual musicians of a group to confess to those times when they have left out a few notes. These small details require great concentration on the part of the producer and a substantial amount of preparation is necessary to be fully ready for the recording session.

Hearing the artist play the pieces before the recording can be very informative. If possible, attending a concert or a rehearsal can help greatly in preparing for the challenges of the recording session.

13.5 Running a Recording Session

In orchestral recording, the producer is normally tasked with directing the recording session. A basic plan of approach is discussed with the conductor ahead of time, and usually each piece or movement of a symphony will be played several times, after which some smaller passages might be repeated as needed, to cover any sections that were not performed well in any of the complete takes. Some conductors like to record a full movement, then take a break and listen. In this case, everyone involved can experience an overview of the work, and the producer and conductor can use the break time to evaluate the complete performance and discuss any changes to the recording plan. After that they run through again, stopping whenever they hear a problem and restarting with some overlapping material to fix the measures in question.

The producer will be juggling many tasks during the course of the session, but the one underlying concern will be finishing the required amount of material before the session ends. Constantly checking the clock, the producer has to prioritize their list of last-minute fixes to be recorded in the remaining minutes, saving the least important fixes for the very end of the session. Any technical details in the performance will need to be documented and reported back to the musicians, with an appropriate amount of diplomacy. If they engage in copious notetaking during the session, afterwards the producer will be able to confidently tell the conductor and musicians which are the best segments of all the recorded takes, so that these can be combined in the editing process to create a satisfactory final version.

For chamber music or soloist projects which tend to run "off the clock", the producer may defer to the artist(s) in terms of directing the flow of events, while suggesting a playback or a break from time to time. In this case, their role is more focused on making sure all of the material is properly covered across the takes, and that a perfect performance can be edited together based on the series of takes performed. There may be more input expected from the producer between takes, when time is less of a factor. At a certain point the artists may ask what measures of a piece are not "covered", that is, which segments were not played perfectly in any take. The producer will need to be ready to report this, based on their notes made in the score throughout the earlier takes. Problematic areas that were never performed well might be notated on a pad, to avoid having to flip through an entire score looking for these spots while the artists wait on stage for direction.

13.6 Postproduction

After the recording sessions are complete, the producer is tasked with creating an edit plan based on notes from the session and comments from the artist. This can be done in several ways, but most commonly markings are placed in the score showing the editor which performance or "take" to use at which time. Some artists like to map out the edit plan themselves, but normally the producer is expected to put together the first edit, with the artist's best interests in mind. In most cases, the marked score is then given to an editing engineer, who will realize the edit following the producer's map in the score.

In the case of a multi-microphone recording that needs refined balancing, a mix session will follow the editing phase. The producer and engineer will normally meet and mix together as a team, although it is assumed that the producer will direct most of the balance decisions based on musical considerations. At this point the producer is responsible for obtaining the artist's approval by incorporating edit and mix revisions requested by the artists, based on the various discussions with the conductor, soloist and so forth.

13.7 Chapter Summary

- The producer is normally responsible for overseeing all stages of the recording process – including the performance and the overall sound.
- A producer needs to have a comprehensive understanding of both music and recorded sound.
- The producer runs the recording session and keeps track of all the recorded takes of each piece, plans the editing map and directs the mixing session.

Chapter 14

How to Carry Out a Successful Recording Session

The key to a successful recording session or live concert capture is preparation. A certain amount of "healthy paranoia" can help the engineer in speculating as to what might possibly go wrong and in assessing which links in the audio chain are the most prone to failure. In the following pages it will be made clear exactly how much preparation is required, and that in some cases "over-preparing" may in fact adversely affect the overall efficiency of the operation. Surveying the hall, troubleshooting the setup, scratch-testing microphones and playbacks with the artist are all covered in this chapter.

It should be evident from the previous chapter that during preproduction, close communication with the producer is mandatory. For projects where a producer is involved, they will most likely be tasked with asking the ensemble for details in orchestration and so on, which will guide the creation of the input list. Contacting the orchestra's administrative staff early on is really the first step in planning a project. The musical score should be consulted for each piece as to the instrumentation, and a discussion should ensue regarding the stage position of each instrument and their corresponding section. The ensemble will be more comfortable with their stage setup during the recording session if they have had the opportunity to rehearse in the configuration proposed for the recording.

Finding a contact at the hall with some technical expertise is also very helpful. Many halls have a certain amount of recording equipment on site, and it may be possible to use some or all of it with the appropriate permissions in place ahead of time. As mentioned in Chapter 5, time should be set aside to visit the hall well before the dates of the recording session. This survey visit can be used to decide on a useable listening room, to check on the length of cable runs between the stage and control room and to ask about loading in equipment, parking and so forth. The input list should include as much information as possible in order to save time at the hall – allocating channels for red light, talkback speaker and a listen microphone on the conductor should be done ahead of time. An input list template can be found in the online resources for this book (see https://routledgetextbooks.com/textbooks/9781138854543), and an example is shown in Table 14.1.

14.1 Equipment, Backups and Options on Location

The first time a portable recording system is used, and before heading out to the hall or recording space, the recording equipment or "kit" should be assembled and tested. This means a microphone with its intended preamp should be connected into the recorder and monitored on headphones. All inputs and outputs of the recorder should be checked, through the mixing console if one is to be used, and testing should include all of the sections of cabling that will be used on location to connect audio from the stage to the listening room, whether the plan is analog or digital, including

DOI: 10.4324/9781003319429-18

Table 14.1 Example of an Input List Showing Input Number, Instrument, Microphone Choice, Etc.

INPUT LIST
Title: Repertoire and Artist Names
Venue: Name of Hall, City
Dates: Recording Dates

Stg Box	Instrument	Mic	Preamp	A to D	Recorder	Comments
H1	Main L	dpa 4006TL	Jensen 1	Prism1	1	hanging, Jensen in Roof
H2	Main R	dpa 4006TL	Jensen 2	Prism2	2	hanging, Jensen in Roof
H3	Aud L	4006	API 1	Prism3	3	hanging, API in Roof
H4	Aud R	4006	API 2	Prism4	4	hanging, API in Roof
F01	Vln I	km 140	Yamaha 01	Yam1/2	CH 9/10	submix buss 1/2
F02	Vln II	km 140	Yamaha 02	Yam1/2	CH 9/10	submix buss 1/2
F03	Vla	km 140	Yamaha 03	Yam1/2	CH 9/10	submix buss 1/2
F04	Vcl	km 140	Yamaha 04	Yam1/2	CH 9/10	submix buss 1/2
F05	Cb	TLM 170	Millennia 1	Prism 5	5	low stand
F06	Harp	MK 4	Millennia 2	Prism 6	6	
F07	Celeste	MK 4	Millennia 3	Prism 7	7	low stand
R01	Perc 1	MK 4	True 1	Yam 7/8	CH15/16	
R02	Perc 2	MK 4	True 2	Yam 7/8	CH15/16	
R03	Perc 3	MK 4	True 3	Yam 7/8	CH15/16	
R04	Horn	TLM 170	Yamaha 05	Yam 3/4	CH 13/14	submix buss 3/4
R05	Trumpet	TLM 170	Yamaha 06	Yam 3/4	CH 13/14	submix buss 3/4
R06	Low Brass	TLM 170	Yamaha 07	Yam 3/4	CH 13/14	submix buss 3/4
R07	Timpani	MK 4	Millennia 4	Prism 8	8	
F08	Flute	MK 4	Yamaha 09	Yam 5/6	CH11/12	submix buss 5/6
F09	Oboe	MK 4	Yamaha 10	Yam 5/6	CH11/12	submix buss 5/6
F10	Clarinet	MK 4	Yamaha 11	Yam 5/6	CH11/12	submix buss 5/6
F11	Bassoon	MK 4	Yamaha 12	Yam 5/6	CH11/12	submix buss 5/6
F12	Conductor	KMS 105	Yamaha 16			for patch session
F13	Red Light					for patch session
F14	Telephone					for patch session
F15	Talkback	AT mic	Millennia 4			for patch session

H = Hanging from ceiling F = front stagebox R = rear stagebox
front stage box under conductor's stand, rear stage box in front of Timpani

optical, coax or ethernet cable. Opening up a live microphone with loudspeakers in the same room is not recommended, especially when testing omnidirectional polar patterns. Even the simplest devices should be checked, such as the producer's talkback microphone and the associated talkback loudspeaker that will be placed on stage. A great deal of time can be wasted working out these small details at the hall if they aren't properly prepared beforehand. The loudspeakers should be tested in conjunction with the intended volume controller and tested by playing back some good-quality multitrack audio directly from the recording device, simulating a playback at the recording session.

Hard-drive formatting and testing should also be done in the days before the recording. Not every hard drive is compatible with every computer, so new drives should be formatted and tested by arming all tracks and recording for a few minutes, using the same computer and DAW software that will be used at the recording session. The recommended format for audio recording on hard drives is exFAT, because it can be easily read by either Mac- or PC-hosted workstations. This will eliminate the need to copy the media between file systems such as APFS (Mac) and NTFS (PC) if a project needs to be moved between DAWs in the postproduction stage.

Plan on reserving an entire day for this, so that the process is not rushed and that a decent amount of time can be spent on properly organizing and labeling all the gear, making sure enough "spare parts" are included (cables, adaptors, etc.). If at all possible, any rented equipment should be picked up a day early and tested in combination with the rest of the kit. This will ensure that there is compatibility within the system and that all the necessary interconnecting cables are on hand. A complete list should be made of everything that is needed, down to flashlight batteries, pens and pencils and gaffer's tape (don't ever assume the hall will provide tape). It is also a good idea to test how the equipment will be transported. Loading a vehicle in different ways to find the best fit can be done ahead of time, and in some cases a simple diagram might be created to document the best "load". An equipment checklist template is available for download among the book's online resources (see https://routledgetextbooks.com/textbooks/9781138854543). An example checklist might look like this (Table 14.2):

According to the Merriam-Webster.com dictionary, Murphy's Law states that "anything that can go wrong, will go wrong" [1] – according to this logic, the only time a backup recorder will be really needed is when there isn't one running. Of course, a better line of reasoning is that once in a while the master recorder will fail, and since it cannot be predetermined when that might be, it is necessary to have a backup system on every recording session.

A backup recording system is absolutely mandatory and can save the operator from themselves as well as from technical problems. Accidentally dropping out of record is a very common issue when using computers as recording devices. The fragile tendencies of both computers and hard drives range from temporary "freezes" and hiccups to full-on "crashes", and these may occur with or without the help of the operator. Simple tasks such as typing a name into a marker window can stop the recording if the engineer isn't completely paying attention. Even chamber music sessions running without a union time clock should be protected with a backup, since the artists should never be asked to repeat a once-in-a-lifetime performance because of a technical problem.

If the recording is planned as a multitrack project and a second multitrack is a complete impossibility, then at least a separate two-track recording of the live mix should be sent to a second device. This is not ideal, however, if the monitor mix is actually being generated by the main recording system, and somehow that system crashes – but at least a two-track backup of the live mix will help in the case of a dropout or a missed downbeat. If a four- or eight-track system is available, consider keeping some elements separated, such as the orchestra mix and the soloist microphones on a concerto recording, so that some effective rebalancing will be possible after the fact if the backup is actually needed.

Many engineers say they run a backup recorder for "redundancy" – a word that is commonly defined as referring to something that is "unnecessary". One always hopes that the backup recording will not be needed, but it is certainly a necessity for orchestral recording and for live concert recordings of any ensemble type. A backup recording system will help protect against technical issues and the more common problem: human error.

Table 14.2 Example of an Equipment Checklist for a Specific Project

EQUIPMENT CHECKLIST

✓ **STAGE:**

microphone preamps: 8ch Millennia, 4ch API, 2ch Jensen (in case with APIs)
mics/clips: 3 x 4006TL with silver grids 4 x KM140 11 x MK4 4 x TLM 170
2 tall mic stands (14') 18 reg. mic stands w/ booms just in case
2 short stands + booms (kick drum type)
short mic cables - ten?
longer mic cables – 24
4 x 8ch XLR snakes 2m/6'
28ch snake at 30m/100' long
1 x 20m/60' 8ch snake XLR for connecting to ceiling lines
small powered talkback speaker, red light w/ power supply, telephones
Gaffer's tape/artist tape/electrical tape/zip ties (tie wraps)
measuring stick or extra tall stand/laser measuring device
video camera
Microphone hangers and fishing line
5' Stereo bar for hanging main microphones
Power strips (2) and extension cords, short and very long (30m/100')
2 x 30m/100' video cable

✓ **CONTROL ROOM:**

8ch A to D (Prism?) in same case as 8ch Millennia and Tascam 24trk
Euphonix MADI to AES converter (need 16 channels at 88.2KHz
Yamaha DM 1000 console w/ hi res AES cards + 2 Yamaha AES harnesses
Genelec 8050s/stands
small video monitor
16ch Fem XLR to TRS for Tascam inputs
adapter kit (tackle box)
3 x 8ch XLR snakes 2m/6'
8 x short mic cables
headphones
talkback mic with momentary switch
UPS/Battery backup, power conditioner
Laptop computer with Pyramix software and RME MADI interface
2nd Laptop (Backup)
hard drives for both computers and backup
power strips (4) and AC cables (3 x 8m/25')
MADI coax/word clock cables (5)
office supply kit (Pens, pencils, sharpies, pads of paper, batteries)
small tool kit (screwdrivers, wrenches, pliers, cutters, soldering iron + solder, flashlight)
more gaffers tape and artit/paper tape
small table lamp

Part of the preparation time should be spent considering what will be done in the event of some elements of the recording chain failing. Again, a certain amount of paranoia will help the engineer configure a solid plan B for every scenario, especially for those recordings in remote locations. For instance, what might be done if an eight-channel unit of high-quality microphone preamps fails? Is there some suitable alternative available as a backup? Does it have the same type of outputs, or will it require a different breakout cable? Are there eight decent preamps in a mixing console that can be used? If so, can the microphone signals be easily received from the stage on analog lines? What other cables might be needed to reconnect in this way? These are all questions that should be answered well before the recording session, so that in the event of any technical trouble, changes can be implemented in the shortest amount of time. For small ensemble recordings in extremely remote destinations, it might be worth asking the producer what brand and type of laptop they are bringing, in case it becomes the replacement recorder, and to bring different connecting cables and hardware drivers that will allow a recording interface to communicate to some other computer than the intended device. Having the recording software installer available on a portable drive to load onto any computer is also a good safety measure.

14.2 At the Hall

Upon arriving at the hall for setup, it is a good idea to ask about the rules. With unionized stage crews, the rules tend to be similar from place to place, while each hall and its union local might follow a slightly different interpretation of the rules. Whether or not the stage crew is union, care should be taken not to do anything on stage that isn't preauthorized – for example, moving the podium or adjusting instruments such as piano (including the lid) and celeste – and any union breaks should be respected. It is common that during a meal break the stage is off-limits, but work may continue in the control room. Some halls can be very relaxed, and if the recording team are careful to ask in advance, they might be able to have stage access while the crew is on break or before an official call time. By sharp contrast, there are a few venues with nonunionized crews that follow such strict rules one wonders how anything gets done.

Packing and load-in should be planned with some efficiency, with equipment separated by location – either stage or control room. Once the setup begins, the input list should be available at both places, whether on duplicate printouts or on everyone's mobile device – although using multiple smartphones generally leads to several versions of the list, unless the entire crew is looking at a live doc online. One printout on the conductor's podium and one in the control room near the recorder is usually best. This way, updates can be done to each in red pen, and there is less confusion. As soon as a "final" version of the list exists, usually after the recording is finished, the original file can be updated and saved on the same hard drive as the audio files, or on a shared space online. It is also good to save drawings and pictures of the stage setup and control room for the next time the team is back at the same hall.

For the best possible signal quality, the microphone preamps should be placed as close to the source as possible – that is, on stage, or for hanging microphones, in the ceiling. This way, the conversion to line level signal can be done early in the chain, preserving the fidelity of the microphone signal [2]. If the preamps are part of the mixing console needed in the control room, a location close to the stage should be chosen for the listening room in order to minimize the loss of quality. The stage crew should be consulted as to where the preamps might be "installed" during the recording – a place where they aren't in the way, but where they can also be easily reached and operated.

Once the control room is configured, the monitoring environment should be evaluated by playing a few excerpts from some favorite recordings – see Chapter 2, "How to Listen", specifically Section 2.3, for more details. The producer and engineer should listen together and discuss the characteristics of the control room so that they may then relate the sound and balance of the upcoming recording session to an established assessment of the listening room. The same applies here for a recording studio session – the control room should be evaluated by playing back some familiar program. Once the listening environment is felt to be at least "workable", a suitable reference recording can be assessed that is directly related to the particular project at hand. The producer will normally have a reference they will want to hear as well. With all references, it is important to note that the comparison is meant to be "rough", and that an exact sonic replica of the reference is rarely the intention – rather, the new recording should be "in the ballpark", or somewhat similar to the reference.

Microphones should be installed on stands with adequate cable free of tape to allow for changes in the angle, and to allow for full extension of the boom. A good amount of cable should be left at the base of the stand so that the microphone can be easily repositioned as needed. The main microphones and outriggers will need enough slack to be raised up several meters to their intended height, as well as enough extra cable to be moved around as necessary. Some nominal level should be set as a starting point, such as 25–35 dB on all preamps, so that at least some signal will be present on all channels and will show a decent amount of meter deflection on the recording device.

14.3 Troubleshooting

One very important skill that should be mastered before heading out on location is the ability to quickly troubleshoot and solve technical issues. It can be very stressful recording live concerts or sessions with a union time clock running. There is no time for panic when something goes wrong; the problem must simply be fixed as quickly as possible. Two very practical skills come into play here – troubleshooting and planning ahead. Of course, nothing replaces proper preparation and planning for equipment breakdowns, but the engineer needs to able to quickly assess what part of the recording chain is causing the problem so it can be addressed directly. A proper plan for a recording session includes solutions to any issue that might be encountered. It helps to be somewhat paranoid here, assuming the worst, so that any issue can be overcome, and the recording will still be successful. Here are some examples:

Problem	*Possible Solution*
One of the main microphones is defective	Bring an identical spare, or extra pair
Less important microphone is defective	Bring extras
Computer crashes (fatal)	Have second computer available (laptop)
Microphone preamp not working	Extra preamp, or use console preamps

The list can be extensive – but what is important is that every possible scenario has been considered ahead of time, and there is a plan in place for each potential problem so that no time is wasted in discussing what should be done.

Troubleshooting audio issues is a very simple and logical process. Someone new to recording might assume that problems during a recording session are best solved by methodically replacing elements of the recording chain from the beginning to the end, starting with the microphone, then the cable, and so on. In fact, the best place to start is somewhere in the middle of the chain. This way a quick test will indicate if the problem originates before or after the point which was diagnosed, and by starting somewhere in the middle, half of the recording chain can be quickly eliminated, which efficiently narrows down the location of the problem.

Here is the *golden rule* of troubleshooting – at any given point in the setup, swap the problematic channel with an adjacent channel that is working normally, and observe the outcome (see Figure 14.1):

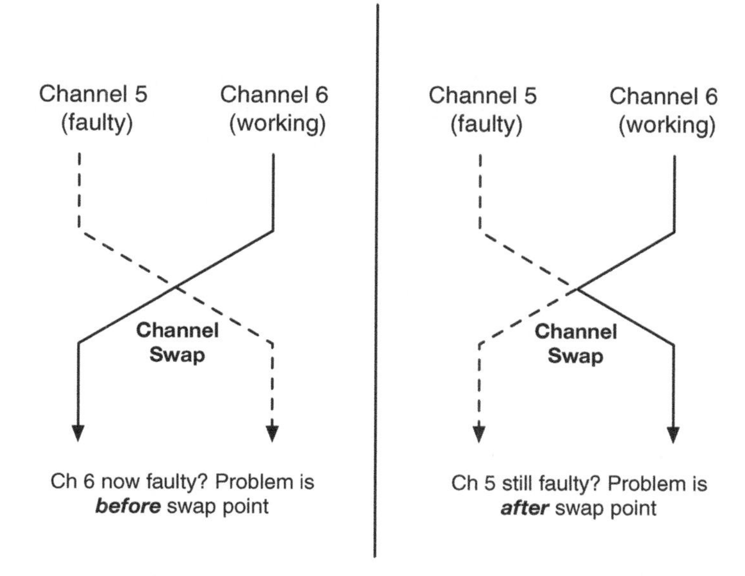

Figure 14.1 Swapping Channels as a Troubleshooting Tool

- If the problem follows the swap, it resides "upstream", or earlier in the chain.
- If it stays the same, it resides "downstream", or somewhere after the point where the connections were swapped.
- *Very important:* Once the diagnosis has been performed, the two channels should be immediately "un-swapped", returning them to their original state, to avoid further confusion.

It may seem obvious, but it should be pointed out that by swapping a bad channel with good channel, we are testing two parts of the chain at once. By sending the confirmed working signal to the second half of the faulty channel's chain, we test the validity of the faulty channel from that point forward. Also, by sending the first half of the faulty signal chain down a path that is known to be working, we can confirm whether or not the problematic channel is at fault, up to that point.

EXERCISE 14.1 USING CHANNEL SWAPPING AS A TROUBLESHOOTING METHOD

Channel swapping can be performed by either physically repatching a connection or changing the routing in the computer software or digital interface. Pairs of stereo channels are easiest, since the signals end up in separate left and right loudspeakers, so the diagnosis is immediately clear in the listening room.

1. As an example, let's consider a noisy input from the stage, say "channel 5", which happens to be the left of two microphones on a piano; the preamp is also on stage, and the microphone signal is heard in the left loudspeaker in the control room, after having

run through the recording computer. So, at the point the signals enter the control room, channels 5 and 6 are swapped.

2. If the noise jumps to the right side, the problem is coming from the stage, and the next logical place to swap the lines would be the microphone cables into the preamp channels. If the noise remains on the left, however, the buzz is originating in the control room.
3. Another example might be a "dead" line, channel 5 in this example. Once again, the "channel 5" stage line is swapped with a working line – if the dead channel follows the swap, we can look to see if phantom power is off or the microphone hasn't been properly connected on stage. If the dead channel stays in the same place, the computer can then be checked to see if channel 5 is muted, or not in an "input-monitoring" or "record-ready" state.

Above all, it is important to remain calm while thinking through the troubleshooting process, especially if it takes place during the recording as opposed to the setup, when there is more time to address technical issues. Acting professionally during troubleshooting will help to gain the confidence of the artists. The goal is to look calm but concerned, even in the most stressful moments of a session – a successful engineer must learn the art of "panicking on the inside".

14.4 The Scratch Test

With all the microphones connected and all devices turned on, it is time to check all the lines. A "scratch" test should be performed on every input – lightly scratching the protective screen in front of each capsule will generate a substantial signal on that particular channel without showing much meter deflection on any adjacent channels. Scratching, as opposed to tapping, will excite more high-frequency content that can help indicate that the microphone is working properly.

EXERCISE 14.2 PERFORMING A SCRATCH TEST

This is done onstage or in the recording studio:

1. While one person scratches each microphone, stating the name of the channel (e.g., first violin, or piano left), the signal is verified in the control room that it is present on the correct channel of the recorder, and clearly heard when that channel is in "solo" mode, with all other input channels cut.
2. Once the input is confirmed, it should be quickly auditioned without scratching to be sure the line is noise-free.
3. Main microphones should be scratched at a reachable height before they are raised up to their intended heights, at which scratch testing is no longer possible.
4. For electronic instruments using a direct input, such as a keyboard or electric bass, a simple dynamic microphone can be quickly connected to check the line to the control room, rather than waiting for the instrument to arrive.

Noisy or "dead" channels will need troubleshooting, and the previous section (Section 14.3) provides a full description of that process. Once everything is working properly, the mains can be raised up to their intended starting height, and cables on the stage deck can be tidied up. It is good practice to keep cable taping to a minimum – if an adjustment is needed in the first few minutes of the session and the microphone stands won't move because of too much tape, time will be wasted undoing the work.

"*If you re-patch, you need to re-scratch*" – any last-minute changes will require those channels in question to be retested. This will confirm that any channels that were temporarily reversed during troubleshooting have been returned to their original patch points, and that phantom power is active where needed. Forgetting to activate phantom power after a quick microphone change is an all-too-common mistake, but it can be easily discovered and corrected by consistent scratch testing.

Just before the session begins, while the musicians are getting ready in their seats and warming up, the positions of the spot microphones need to be checked. Any microphone can be moved, and with a small amount of boom work, compromises can be achieved so that both the musicians and the recording team can be satisfied. It is also a very good idea to get on the podium, look around for any stands that might be blocking the musicians' view of the conductor and then adjust them before anyone has time to complain. This should be done as subtly as possible, without drawing too much attention to oneself.

14.5 During the Session

The first moments of the recording session will be dedicated to working on the sound and balance, getting the input gains properly set and establishing a decent monitor mix. This is a time during which great focus is required, and a small amount of nervous energy can help the recording team think quickly and make confident decisions regarding microphone moves and refining the balance. During the first few minutes, the conductor will direct the ensemble to perform some selected passages while waiting to be told when the recording team is ready to record a first take of a movement. The producer and engineer will need to be ready to move ahead with the recording as quickly as possible once they are confident that the microphones are in their optimal positions.

Announcing take numbers: this is also called "slating takes". The producer may want to announce each take number into the hall over the talkback speaker. This helps to keep the recording process running in an organized manner, as the musicians know that when they hear the "slate", the recording system is running and they can begin to play. Others may ask the engineer to take care of the announcements, and some producers prefer that the take numbers are not heard by the musicians, as it might be distracting. In this last case, the slates may be recorded directly to the DAW or simply announced in the control room to ensure the operator and producer always agree on the take number being used. This ensures that the producer's notes and the recording system file numbers will correlate and that the correct musical content is used throughout the edit plan.

False start: this is when the musicians begin to play, and almost immediately stop and start again, for whatever reason. Most of the time there isn't a chance to call a new take number before they start over, especially if the takes are being announced onto the hall or studio. Rather than asking the musicians to wait for a new number, the producer or engineer will call for a "false start" in the control room, and the recording continues without stopping or changing to a new take number. The producer will mark the take in their notes with "f.s.", and the DAW operator should either move the take marker to the correct starting place or add a second marker indicating the false start.

For those readers who might have skipped over the previous chapter, Chapter 13, The Role of the Producer in Classical Music Recording, good communication with the musicians on stage or in the studio is a major concern (see Section 13.3 specifically). Musicians who feel "ignored" by the recording team in the control room will not perform with their full attention or to their full ability, and the resulting recording will reflect this sentiment. Any questions or comments coming from the stage or studio should be immediately addressed. If an important conversation is underway in the control room, the artists should be told that the recording team needs a minute to discuss, and that the artist's concerns will be addressed shortly thereafter. This is simply a question of paying attention and exercising good "studio manners". The "snoop" or listen microphone will come in handy at this point in order to achieve clear two-way communication. This input channel can be opened between each take to clearly hear the conductor talking over the background noise of the musicians chatting and playing, and general stage movement.

14.6 Playback Sessions and Artist Rapport

Playbacks are an efficient use of time during the orchestra breaks, and a specific protocol should be followed so that the artist is comfortable with the process. There is a certain level of skill required in smoothly running the playback while avoiding any awkward moments when the musicians (or representatives from the record company) are in the control room.

EXERCISE 14.3 PREPARING AND RUNNING A PLAYBACK SESSION

Here are 10 important playback guidelines, presented here as an exercise:

1. Before the artist arrives in the control room, ask the producer which take will be heard first, and cue up the playback to the beginning of the music, after any onstage talking, take announcement or false starts.
2. Quickly test that the playback will actually start before they arrive. Make sure the recording system isn't left in "input mode", and that the volume control isn't muted or dimmed.
3. Wait until the artist is seated and with their music open, ready to listen, before starting the playback.
4. In the case of some discussion before the playback begins, look to the producer for a sign as to when to press play.
5. Keep an eye on both the producer and the artist in case there is a request to stop and backup to replay a certain passage.
6. Also keep an eye on the producer at the very beginning of the playback in case they motion for the volume to be adjusted.
7. Pay attention while audio is running – if the artist reacts to something in the performance, but doesn't ask to stop, keep the playback running but add a marker, or make note of a timing, so that the spot can be easily found and replayed on request.

8. Playbacks are not the time to start a conversation in the control room while the artist is trying to listen. This can be distracting and/or interpreted as disrespectful.
9. Before returning to "input mode" and "record ready" on the recorder, it is best to wait until after playback discussions are complete and the artists have exited the control room.
10. Check that the monitor level is dimmed before activating input mode so as to avoid blasting the control room with noisy stage sound as the musicians are returning to their seats and warming up.

In terms of rapport, an effort should be made to team up with the artist, always addressing their various concerns in a supportive and professional manner. Discounting issues raised by the artist that seem trivial or negative will most likely alienate them to some degree. Following the hierarchy laid out in the previous chapter, the engineer should let the producer speak first when an artist has a question, even if that question is directly related to the sound. If the producer also looks to the engineer, however, without speaking, then it may be a sign that the producer wants the engineer to engage and tackle the question directly. Again, "knowing one's place" and acting accordingly is the key, based on the best read of the situation.

By working in this way, the team can avoid giving the artist opposing opinions that may be confusing. As stated in Chapter 13, if the engineer wishes to counter the producer, it should be addressed after the artist has left the control room. Once a particular producer and engineer have worked together a few times, any delicate or awkward moments with the artist can be navigated with ease and fluidity.

14.7 Chapter Summary

- Meet with the producer to discuss instrumentation and microphone placement.
- Visit the hall ahead of time and test all equipment and connections off site before arriving at the hall.
- Take time to practice troubleshooting so that the process is fully understood and becomes second nature.
- Playback sessions should be taken very seriously.
- Always run a backup recording, and don't forget to scratch-test the microphones.
- Include the artist as a member of the recording team, always working with them in addressing their concerns.

References

1. "Murphy's Law." *Merriam-Webster.com*, https://www.merriam-webster.com/dictionary/Murphy%27s %20Law, accessed July 13, 2023.
2. Eargle, J. (2005). *The Microphone Book.* Burlington, MA: Focal Press, Ch. 8 (p. 135).

Chapter 15

Live Concerts, Live Opera and Productions with Video

Live concert recording has its own built-in challenges. There is never an opportunity to ask the conductor to start the concert over once they have walked out on stage and launched the orchestra into the first piece on the program. Input levels, the amount of reverberation and overall timbre will probably differ between the dress rehearsal in an empty hall, and the concert with a full audience and an orchestra that is playing with more energy and focus than in the morning run-through. Singers quite often only "mark" their parts (singing in half-voice) in a dress rehearsal, saving their voices for the concert, unless they agree to the dress being recorded and used as an alternate performance in the editing process.

In live opera performances, we witness the entire orchestra being jammed into a pit that is partly or mostly under the stage, and a few singers meandering over a rather vast area, with a great deal of movement and head-turning while performing. Providing audio support for large-scale video productions can be a logistical nightmare for an engineer. Video directors will ask for all the microphones to be out of all the camera shots (i.e., invisible), and the additional lighting and dimmers add extra noise to the background sound. Some well-practiced advice is offered in this chapter for how best to navigate the world of live recording.

An easy way to document details such as conductor and soloist names, pieces performed, and so on is to scan (or take a picture) of the concert program and include that file on the same hard drives which will be used to record the audio, as well as in a general shared folder of concert programs when recording a concert series (for example, archive recordings of an entire season at a particular festival or hall). Many organizations can provide the program ahead of time as a digital file (PDF), which will save a step.

15.1 Live Concerts

Many of the guidelines in Chapter 14 for running a recording session apply just as well to capturing live concerts. Practices such as visiting the hall ahead of time and organizing and pretesting the equipment are all equally beneficial. Running a backup recorder is an absolute necessity, as there is only one chance to record each piece. In a session, even an orchestra can be asked to replay a short section that might be lost due to technical reasons, or a break can be called to quickly remedy

DOI: 10.4324/9781003319429-19

a problem – but a live concert cannot be stopped once it has been started, so two recorders are needed at all times.

The presence of the audience changes a few aspects – there will be more background noise (coughs, chair movement) and the stage will have to be much tidier visually, in terms of cables and stands. Musicians tend to play differently in concert as opposed to a recording session, where the audience brings some tension to the ensemble as they are expected to deliver. This nervous energy can be a positive influence, but it can also be evidenced through the presence of a certain "tightness" in the way the sound is produced. Soloists and members of the orchestra generally try to project their sound to fill the entire hall, rather than performing for the microphones. This results in an overall sound that can be somewhat rough and "pushed", as opposed to a more refined sound with greater resolution in dynamics, as is more common in a recording session.

Performing a complete symphony from top to bottom can bring great spontaneity and perspective to a piece of music, while a recording session allows for each movement to be repeated, honed and refined, the result being a more concise interpretation of the overall work. Editing between two performances separated by 24 hours can be less effective than two takes of a movement that were recorded in direct succession during a recording session. That being said, most orchestral recording is done live these days because of the prohibitive cost of orchestral recording sessions. The conductor is required to take great care in achieving consistent performances each night from the ensemble, with matching tempos.

One distinct advantage of live concert recording is that the rehearsals, or at least the final or dress rehearsal, will normally be in the same space as the concert, so there is a good opportunity for the recording team to listen over the microphones and make adjustments to their placement before the concert.

One important note – if the orchestra is a union operation, the rehearsal can only be recorded if the orchestra is paid. In most cases, musicians are not paid for a rehearsal recording, only for the concert and possibly a "patch" session immediately after the concert, where a few spots may be reperformed to fix major mistakes, or to record an ending of a piece without applause, or the beginning of a slow movement without audience noise.

In the case of a unionized ensemble, no attempt should be made to make an illegal recording of the rehearsal. Many orchestras will send a representative to the control room during the rehearsals to check that no recording is taking place, and the repercussions of breaking the rules can be dire. It is not worth the risk, as the recording team and the record label involved may be "blacklisted", that is, prohibited from recording any union orchestra or musician in North America.

Hanging microphones: hanging requires much more work than placing microphones on stands and takes extra time, but it improves the overall look of the stage from the house. Once the microphone positions are set, the advantage is that they can usually remain in place through a series of performances. The trick is to be sure the ensemble and its various parts are in the right place on stage each night, so that they are properly located under any hanging microphones. For example, there may be harp in the violin concerto during the first half, but not in the symphony that is performed after the intermission, so it will then be removed. The harp will need to be repositioned in exactly the same place the next evening for the second performance of the concerto, and the engineer will need to work with the stage crew to be sure this happens.

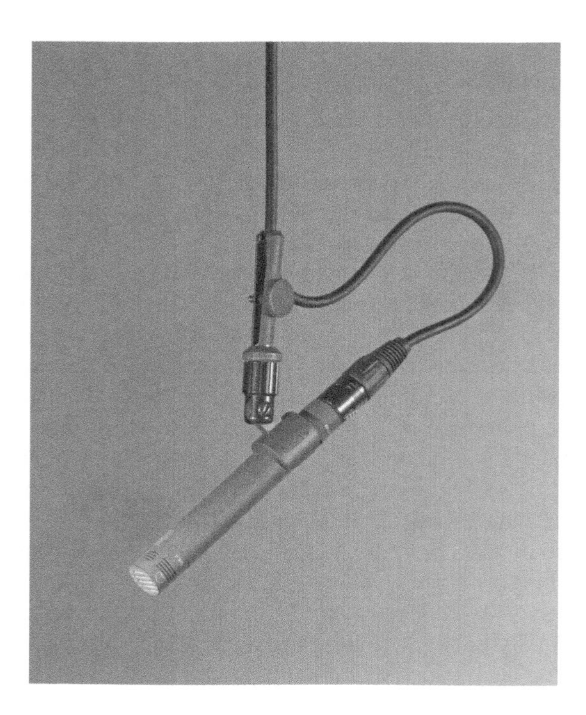

Figure 15.1 A Typical Hanger Design for Attaching a Microphone to Its Cable
Credit: Jack Kelly

Some stage crews that are familiar with hanging microphones on a regular basis may have fishing line and hanging clamps available, and the crew might suggest possible positions around the stage area to tie off the lines. Most stagehands have a wealth of rigging experience and they can be a big help in safely hanging and tying off fishing lines. The recording engineer should be ready to provide all the hardware, so that they are never stuck without the necessary materials. Hanging clamps, fishing line, fasteners such as cable ties, heavy paper clips and electrical tape should be part of any hanging kit used for live concert recording.

At the very least, the front line of microphones should be suspended from above in order to avoid an ugly and potentially dangerous row of microphone stands at the edge of the stage and in full view of the audience. This includes the main system, the wide outrigger pair and possibly soloist microphones. Hanging woodwind lines may or may not be necessary, and at the back of the orchestra it is generally accepted that brass and percussion microphones on stands are not much of a visual distraction. It is a good idea to ask the contact at the hall what is normally done when it comes to hanging microphones, and to plan accordingly. Hangers or hanging clamps will accommodate a regular microphone clip that allows for varying the vertical and horizontal angle of the microphone (Figure 15.1). Hangers that clamp onto the microphone cable can be easily adjusted in height using a ladder from the stage floor, if a small amount of cable slack is left. This alleviates a trip up into the roof for each minor alteration. Clamps can be homemade, although these days a range of suitable hangers is available at affordable prices.

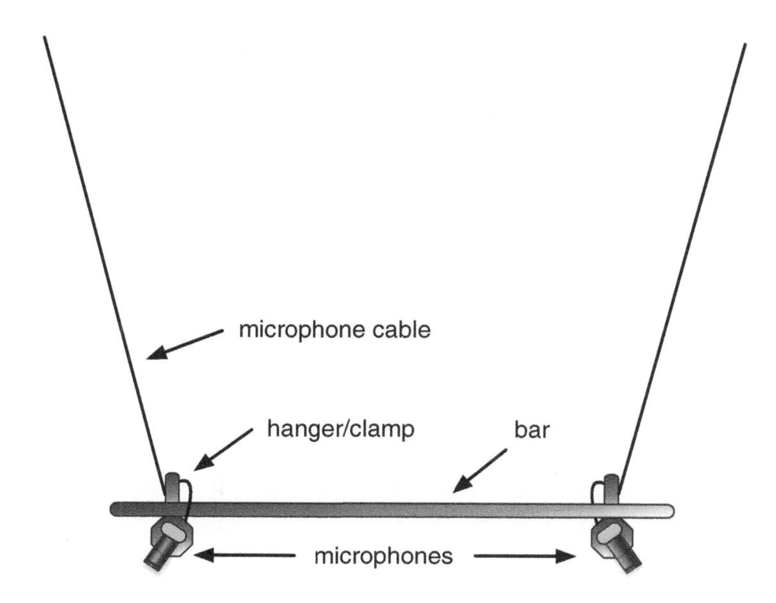

Figure 15.2 Diagram of Hanging Main Microphone System

EXERCISE 15.1 HANGING THE MAIN MICROPHONE SYSTEM

1. For the main microphones, use a lightweight bar with clamps or a small pipe to keep the system rigid – it will then be more stable when the system needs to be pulled forward or backward to the desired position (see Figure 15.2). For a homemade bar, small clamps can be added that will accept the thread of a microphone clip.
2. Drop in the cables from the ceiling to a height where they can be easily reached from the stage floor (or all the way to the floor), and then attach the bar and the microphones. Cables should be attached to the bar with both cable tie fasteners and electrical tape, as the tape stops the microphone cable from slipping under the fastener, while the fastener supports most of the weight or load.
3. For a less obtrusive look, a short length of fishing line can be attached between the left and right main microphone cables just above the individual hangers. Fishing line can also be used when a bar of sufficient length is unavailable. Fasten paperclips to each end of the line so that it can be quickly attached and detached, maintaining a fixed spacing for the microphones.
4. Choose drop points from the ceiling that have a wider spacing than needed and pull them in to the width of the bar or the alternate fishing line. This will hang evenly in place without swinging around.
5. Two additional fishing lines can be attached to the two clamps on the bar to pull the system toward the stage or back into the hall, as needed. These lines should also be tied off at a wider spacing than the bar for added stability. Again, electrical tape is the best fastener for securing these lines.

Microphone heights of some sections may need to be slightly higher when hanging, to avoid accidental contact as musicians and stage crew enter and leave. String spots, for instance, may be 10 cm or a few inches higher than a typical stand height, but then the capsule can be faced straight down to help attenuate leakage from adjacent sections.

In most instances, it is preferable to listen to a rehearsal with the main microphones on stands so that very fine adjustments can be made during the rehearsal, in any direction. This may seem luxurious, but if the microphones are hung in place before any listening is done, there will be a limited number of adjustments that can be made and limited time in which to make the changes (a break in the rehearsal will allow time for one change, then one final adjustment after the rehearsal). The net result will be a less precise placement of the overall pickup, which may have to be compensated for with the spot microphones. If there is an opportunity to have microphones on stands for the rehearsal, the recording team will have to ask for stage access before the concert to get all the microphones into their hanging positions. All of this should be planned out beforehand with the stage crew and according to the hall schedule, to be sure there is ample time available.

A word on the ever-popular gaffer tape or stage tape – it is not that sticky! It is designed for temporary applications and as such, it should not be expected to hold any amount of weight for extended periods, such as an entire show or a series of concerts. Proper fasteners such cable ties/zip ties and electrical tape should be used for hanging microphones and tying off fishing line.

Microphone preamplifiers: as stated in Section 14.2, it is good practice to keep the distance from the microphone to the preamp as short as possible. To follow this same practice when the microphones are hanging, the preamps need to be installed in the ceiling, near the point where the cables are "dropped in". The actual microphone cables can be run up to the ceiling patch points even for the rehearsal when stands are being used – this way the entire chain is tested, and less work is needed when hanging the microphones after the rehearsal and before the concert. A nominal starting gain should be set for all the preamps, including those in the ceiling – around 25 to 35 dB, depending on the microphones used and the dynamic range of the repertoire.

As an alternative, connecting the main microphones to preamps on stage during the rehearsal will allow for easy gain adjustment, but when the microphones are hung and the preamps are moved up to the ceiling the lines should be carefully checked afterward for noise, since the preamp will be connected to a different power source that may create a hum or buzz on the line. It is much better to have the preamps installed in the roof so that there is less change in the signal paths after the rehearsal. In most halls, a member of the crew should be able to access the preamp in the ceiling to adjust the gain while the rehearsal is in progress.

15.2 Live Opera

Capturing a live opera presentation can be a huge challenge that will require a great deal of planning. In order to effectively perform a decent live mix, the stage movement will need to be documented in the score. Quite often the producer or a designated score reader will attend several rehearsals before the dress rehearsal, and will have this information clearly notated in their score so they can call out the action as it unfolds. Unfortunately, footsteps, thumps and props being moved around are all part of the activity on stage. For opera recordings without video, these noises can be highly distracting – a duet during a swordfight will be a complete loss on an audio recording. Fortunately, swordfights in opera are generally kept quite short, and the singing usually only resumes *after* the lethal blow has been dealt.

Normally the pit capture will be quite simple, as the orchestra is cramped into a small space and there is little that can be done to isolate the various sections of the ensemble. The stage sound, however, will not be so condensed, and since the action unfolds in various parts of the stage, proper coverage of the singers and chorus will present a major challenge. A main system hanging somewhere over the pit and stage will provide an overall picture of orchestra and singers, albeit somewhat diffuse, especially for any upstage action. A row of cardioid or hypercardioid microphones located in a low position at the pit edge across the front of the stage can help refine and focus vocal balances. See Figure 15.3 for an example of a typical stage plot for live opera capture, with highlighted microphone positions.

> A good analogy for balancing live opera is mixing audio for a televised ice-hockey game – the audio mixer learns to follow the puck, always favoring those microphones closest to the action, while keeping all other microphones lower in level in an effort to reduce comb-filtering effects, background/crowd noise and the buildup of indirect sound in the mix.

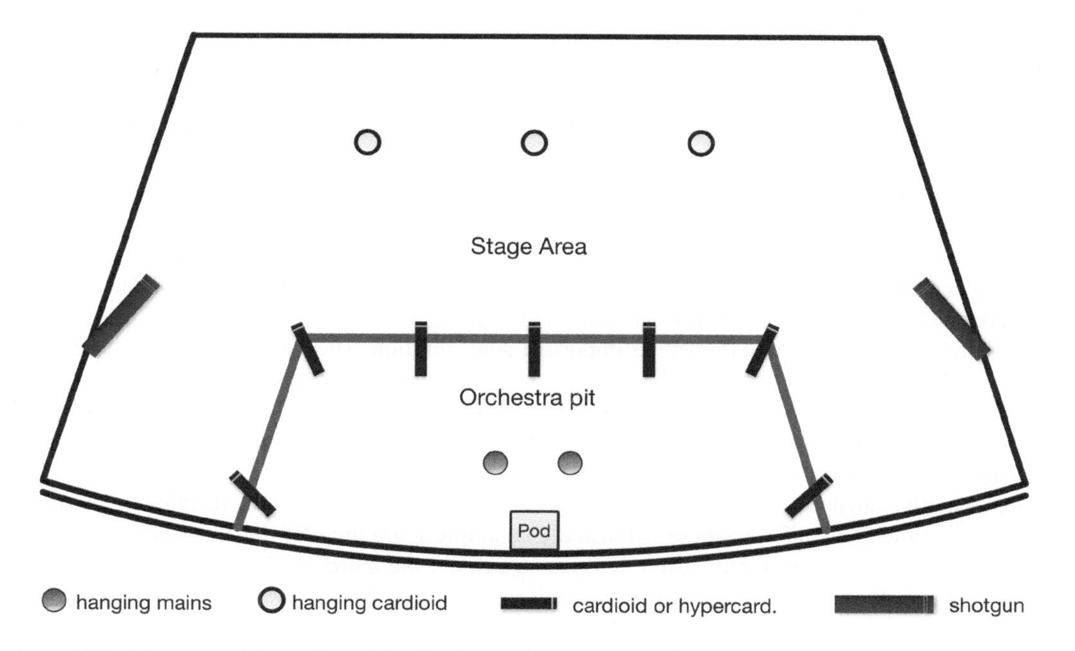

Figure 15.3 Diagram of Typical Stage Plot for Stage Capture in Live Opera

One of the most objectionable sounds in live opera mixes is the effect of too many stage microphones open at the same time, with few of them properly capturing the singer or singers. This buildup of diffuse, off-axis signals will blur the mix resulting in a very roomy stage sound, so the number of open microphones needs to be monitored and adjusted as the singers move around the stage during each scene. Clarity in the voices is critical, and the text should be well understood. Shotgun microphones have good reach and can help with the capture of any singing upstage (farthest from the audience), the resulting sound capture will be less even as compared to microphones with wider polar patterns.

The best combination, if at all possible, is to combine an overall capture using various omnidirectional and cardioid microphones with a wireless "lavalier" or body microphone on each of the lead singers. Signal from the lavalier microphones will be very close and dry, and as such these signals will need to be combined with other channels at conservative levels so that the perspective remains as realistic as possible. A big advantage to using lavaliers is that the singers rarely sound "off-mic" as they move in various directions and sing from all over the stage.

For this strategy to be easily implemented, a separate team should be placed in charge of preparing and testing the body packs, and actually helping to attach the microphones and body packs to the singers' costumes. Special belts can be made or purchased when necessary and, in some cases, pockets for the body packs can be sewn directly into the costumes. Managing frequency bands and installing fresh batteries before each performance are also part of the wireless team's responsibilities.

15.3 Productions with Video

When providing audio support for a video production of a live event, a good mandate to follow is that all the microphones should be invisible. Video directors will want to plan their camera shots without worrying about seeing microphones in the image. Rather than waiting for the video team to complain about microphone positions, it is recommended that placements be worked out ahead of time. The resulting microphone positions may not be ideal for the audio capture but will provide for a clutter-free stage. This means the main system and string spot microphones may need to be suspended slightly higher than what is considered ideal, and more directional polar patterns may be required, as part of the compromise. Lower placement heights may have to be used for soloist and woodwind-section microphones, to avoid being seen in "close-up" video shots.

Technical considerations such as the time-code frame rate and video reference should be discussed with the video crew ahead of time and prepared accordingly. More information on audio–video synchronization can be found here [1–3]. Ideally, the audio recorder should be "locked" to a reference signal provided by the video team, although in recent times some projects with lower budgets have begun to forego this rather important consideration. The cameras may be all running freely, and the synchronization is only as accurate as the internal crystal on each device. For long programs such as live concerts, the sync of the audio and video will most likely drift over time and will need to be "re-synced" at the top of each piece in the program, or even more frequently depending on the circumstances.

One positive attribute to multimedia productions is that lighting and video installations require much more time than audio, so there will usually be ample setup time for the audio crew to "hide" microphones and resolve certain aesthetic issues regarding the more sensitive placements. In live opera productions, the lavalier microphones will need to be hidden very well for close-ups, and various locations in the scenery can be used to hide support microphones, such as trees or shrubs.

It is very important in video productions that the perspective of the audio matches the picture. It may require that the recording take on a much more intimate presentation than might be appropriate for an audio-only production. If at all possible, the audio team should ask for a video feed for the audio control room, so that the engineer can watch the "live cut" or the camera switching as performed during the live event. This will provide a sense of the number of long shots as compared to mid shots and close-ups, so an average audio perspective may be realized that closely matches the image. For live mixing of opera, it is helpful to have a long shot on a second screen, so that the stage action can be followed at all times.

Panning of the stage microphones should be kept narrower than normal, so that tight shots in the video will not be paired with disparate audio. Just as dialog in film is normally anchored to the center channel, a singer's voice should not be coming from the far left or right of an audio mix if they are featured in a close shot in the center of the screen. String sections and orchestra sound in general can follow a normal left-to-right image, thereby filling out the lateral soundstage. Solos in the woodwind and brass sections will want to remain fairly close to center, if any close camera shots are being used in a symphonic video production. When altering the location of a source as it appears in the main pickup, great care should be taken to assess any artifacts such as "ghost" images. For example, a clarinet solo with its spot microphone panned close to the center to conform to a tight camera shot might not align very well with the natural image of the clarinet, which might be localized farther to the left in the main system. A simple adjustment of the balance will help to clarify the image for the duration of the solo section. This is yet another example of the compromises that are required of an audio engineer working on a video production, where the picture takes priority over the sound.

15.4 Chapter Summary

- Recording orchestra and opera live in concert is generally a less desirable approach, but it is more cost-effective and is therefore more common than closed recording sessions.
- Expertise in hanging microphones is required for most live concert recording.
- Try to keep microphones hidden when possible, especially for video productions.
- Contact video crew early to coordinate technical details (time-code and sample rates, video reference, schedule, etc.).

References

1. Holman, T. (2010). *Sound for Film and Television*, 3rd Ed. Burlington, MA: Focal Press.
2. Radcliff, J. (1999). *Timecode: A User's Guide*, 3rd Ed. Burlington, MA: Focal Press.
3. Rumsey, F., and McCormick, T. (2014). *Sound and Recording: Applications and Theory*. Burlington, MA: Focal Press.

Part V

Other Classical Music Ensembles

Chapter 16

Recording Wind Symphony and Brass Band

Wind symphony is a large ensemble made up of woodwind and brass instruments, along with timpani, various percussion elements and normally one string bass. Harp, piano and celeste may be included, depending on the repertoire to be performed. Brass bands, in the British tradition, are made up of all brass instruments plus timpani and percussion and occasionally harp, piano and celeste.

16.1 Wind Symphony or Concert Band

This type of ensemble goes by several other names including symphonic band, wind ensemble or wind band. The group is made up of woodwinds, brass, percussion, and double bass. A possible stage setup can be seen in Figure 16.1, although there is really no standard seating arrangement. While an orchestra is made up of two very different sections – a string orchestra in the front and a symphonic "band" of winds and brass in the rear stage area – wind ensembles are comprised of more similar sound sources and therefore generate a more homogenous sound. This means that it is actually easier to achieve a good overall balance of the entire ensemble on a main microphone system than it is for orchestra. In a recording session or concert performance, wind ensemble can be captured using the same methodology and techniques used for orchestra, with very little adjustment.

16.2 Wind Symphony Microphone Placement

For wind symphony, as with orchestra recording, a pair of mains above and behind the conductor combined with a pair of outriggers works very well. Alternatively, three microphones across the front in a straight line as a left/center/right capture is also a good option, as in Figure 16.2. Center might also be replaced with a narrow L/R, as seems best. Various spot microphones can be added to cover each section (flutes, clarinets, saxophones, brass, etc.) and to catch any solo passages from within each group, as is the typical practice used for orchestra. Double bass is typically part of the ensemble and should be captured separately (especially if it is a single player) and carefully incorporated into the mix so that it is clearly represented in it but not overly emphasized. The balance should be comprised of mostly the main system with a small amount of spot microphone level for focus.

DOI: 10.4324/9781003319429-21

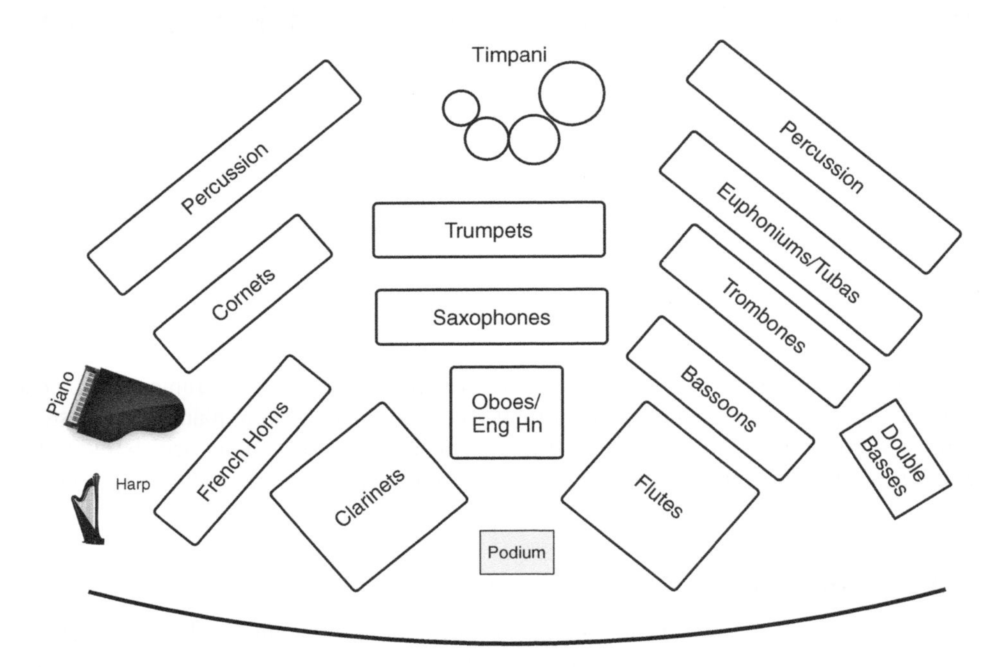

Figure 16.1 Example Stage Layout of a Wind Symphony

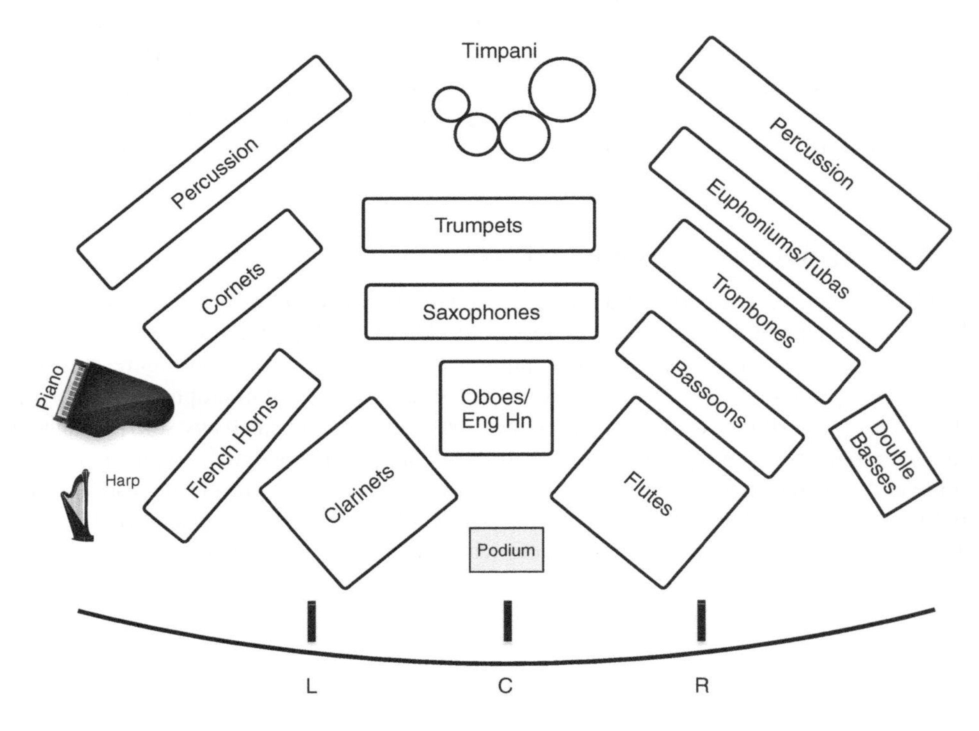

Figure 16.2 Diagram of a Three-Microphone Main System for Wind Symphony Recording

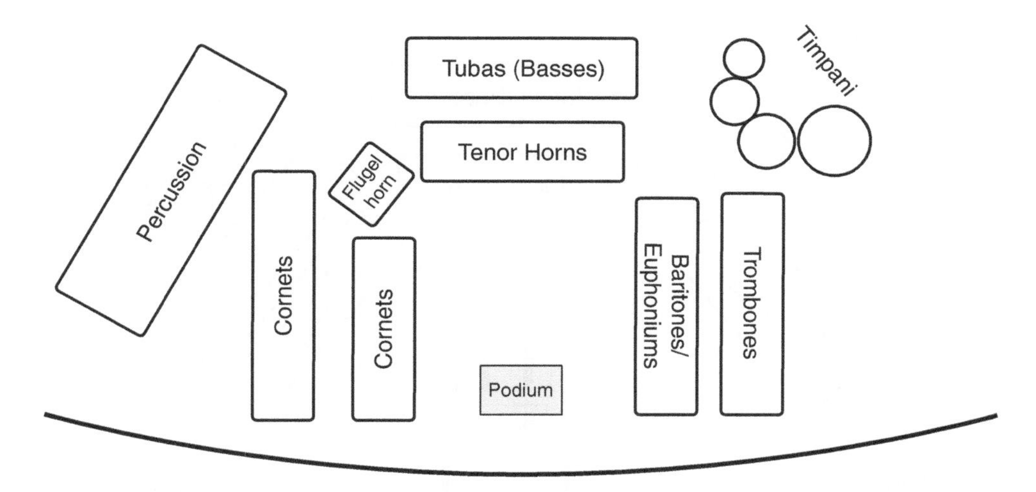

Figure 16.3 Example Stage Layout of a Brass Band

16.3 British Brass Band

The British brass band in the formation we know today first appeared in the early 1800s, and that particular combination of instruments can be found in ensembles throughout the world. British brass bands have a traditional, unique setup. Rather than the standard trumpet, cornets are used for the soprano voices. While the French horn commonly appears in orchestra and wind ensemble brass sections, the tenor horn (or alto horn) is used in the brass band. Flugelhorns, baritones, euphoniums, trombones and both E ♭ and B ♭ tubas (or basses) make up the rest of the ensemble, along with timpani and various percussion. Figure 16.3 shows the traditional layout of a brass band, with the cornet players facing directly across the ensemble.

16.4 Brass Band Microphone Placement

While the wind symphony generally produces sound that propagates toward the audience, the brass band has a more complex array of "directivity". Cornets and trombones tend to produce a direct sound that crosses the stage, while the tenor horns, baritones and euphoniums exhibit an upward propagation. Tuba of course tends to be present everywhere, even though the bell of the instrument usually faces up. The result of all these variables is that the traditional capture point for the main microphone system over the conductor's head will be greatly lacking in clarity from the cornets, and possibly the trombones, as they face sideways on stage rather than playing out into the hall. Additionally, raising and lowering the main system will greatly influence the balance between cornets and horns – a lower position supports the detail of the cornets, and a higher position will round out the cornets while favoring horn presence. Because of this, the best option is to have a main system over the conductor, supplemented with a second main system placed inside the ensemble and much lower in height. This will help to properly capture the direct sound of the cornets, and to some extent, trombone clarity. An example is shown in Figure 16.4, along with the suggested modification of a "V"-shaped ensemble position on stage. This aids in directing the ensemble's energy out into the hall rather than across the stage.

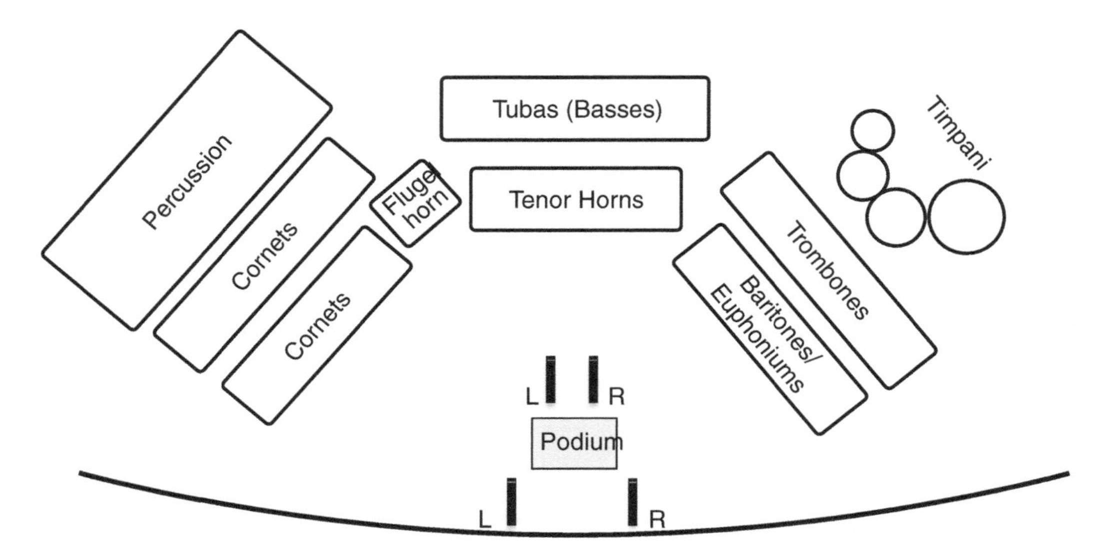

Figure 16.4 Diagram of a Double Main System for Brass Band Recording, and a Modified Stage Position

EXERCISE 16.1 POSITIONING A DUAL MAIN MICROPHONE SYSTEM FOR BRASS BAND

1. Start with a main system at a height of around 3–4 m (or 10–12 ft.), 1–1.5 m wide (3–5 ft.) and placed just behind the conductor's podium.
2. For the closer system, place the microphones slightly in front of the conductor and set height at around 1.5 m (or 5 ft.), at a width of around 30–40 cm (12–16 in.).
3. Set the main system to 0 dB, with all other channels muted.
4. Bring the secondary microphone system into the mix at a much lower level than the main system, as a complementary element (possibly -10 dB relative to the main system, or less).
5. Introduce it slowly, muting it in and out to evaluate the change in balance.

16.5 Chapter Summary

- Wind symphony or symphonic band, along with brass band, will sound darker than an orchestral texture, due to string instruments providing a richer set of harmonics and a stronger presence of high frequencies.
- Brass bands in the British tradition employ the tenor horn instead of French horn, which is more commonly seen in wind symphonies and orchestras.
- With cornets facing across the stage rather than out into the hall, an innovative technique is required for the brass band main capture system.

Chapter 17

Recording Chamber Orchestra

Many chamber orchestras have no conductor – instead they are directed by the concertmaster or the soloist. The Four Seasons (Vivaldi) concerto series is a good example of a piece that requires no conductor. Smaller orchestral groups, ranging in size from 18 to 30 musicians, may need to be recorded using a different approach than that for a full orchestra. The technique of an "expanded tree" is suggested, which treats the ensemble like a large string quartet, providing excellent coverage of the ensemble with minimal use of microphones, allowing for a natural sound and balance.

17.1 Recording Chamber Orchestra Using a Low Channel Count

Quite often, chamber orchestra recordings are realized on small budgets, and it may be necessary to approach the project using very few microphones and a recording system with a limited channel count. In this case, it will be even more important to focus your efforts on the main system, since there will be a very small number of available channels to use for support microphones. Adhering to a multitrack production methodology, it is quite possible to record a small orchestra using as few as eight microphones. Major adjustments in technique will be required, such as the elimination of most string spots, except possibly the bass microphone.

With smaller-sized ensembles, the engineer stands a good chance of capturing almost everything on the main system, since the depth of the ensemble is usually quite shallow. String orchestra can be recorded very well with only a few microphones, using an intimate placement of the main system. Two main microphones might in fact do the job, if there is time available for listening and adjustments of microphone placement- and the positions of the musicians themselves. It is recommended to have at least a bass spot microphone, in order to control low-frequency definition, but the other string spots are not normally needed as long as the main system isn't too far back in its placement. In fact, a normal perspective for chamber orchestra recording will be significantly more intimate than that of a full orchestra, as detail becomes almost more important than blend. Heights for the main microphone system will be somewhat lower, by as much as 60 cm (2 ft.). Other spots can be added as needed for woodwinds and so forth.

17.2 The Expanded "Tree"

A system of three main microphones in a wide spacing can be a powerful tool for chamber orchestra. As a replacement for two main microphones combined with a pair of outriggers as a secondary wider capture, the three microphones of an expanded tree offer the same stage coverage in fewer channels. In fact, it can be thought of as a combination of the Decca tree with a pair of outriggers – the number of microphones is reduced from five to three (see Figure 17.1). For most chamber

DOI: 10.4324/9781003319429-22

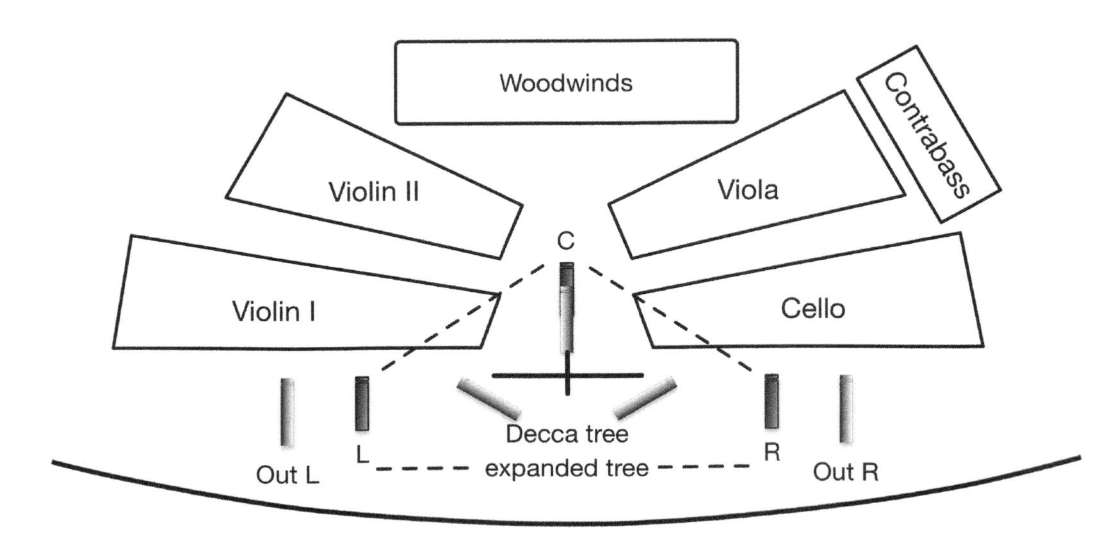

Figure 17.1 Diagram of Decca Tree and Outriggers vs. an Expanded Tree

groups, the width of the ensemble does not warrant the use of a second, wider pair of microphones to cover the smaller forces (i.e., mains and outriggers), yet a typical "main pair" on its own tends to overemphasize the first desks of each string section.

How do you get the stability and coverage of a Decca tree system with outriggers, using fewer microphones? Preserve the center microphone of the tree, move the left and right main microphones out toward the normal position of the outriggers and eliminate the outrigger pair. The result is an "expanded tree" of only three channels.

An expanded tree offers strong center, and the three-channel system works very well for surround-sound projects, providing discrete audio capture across the left/center/right front channels. The wide spacing of the left and right microphones enhances the section blend and provides an increased low-frequency directivity normally associated with an outrigger system. A spacing of up to 3–4 m (10–12 ft.) is possible, and the center main microphone can be placed slightly forward into the group for a more intimate capture of instruments in the middle of the ensemble. Using separate stands for each of the three microphones is necessary because of the very wide placement. Unlike the Decca tree, the center channel is normally set to the same level as left and right because of the large spacing between the microphones. An entire chamber group might be recorded using only eight channels, as follows:

- CH 1/2/3 Expanded tree (left/center/right)
- CH 4 Bass
- CH 5/6 Left/Right Woodwinds and Brass (stereo pair)
- CH 7 Harp
- CH 8 Timpani

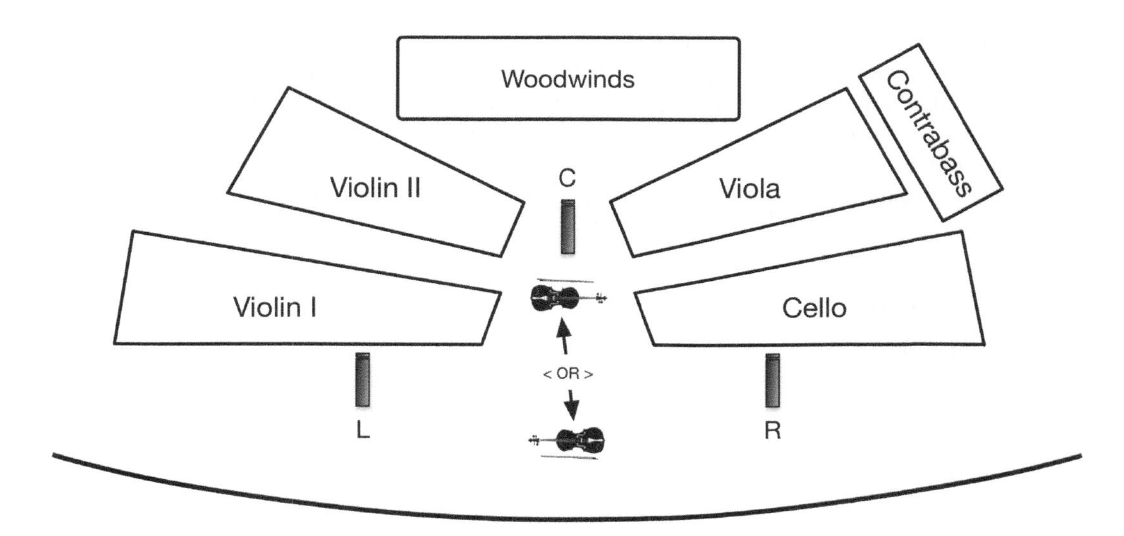

Figure 17.2 Diagram of Chamber Orchestra with Optional Positions for the Soloist

In the above example, the woodwind/brass capture should be placed so that it accurately and evenly supports the back half of the ensemble, possibly including percussion as well.

17.3 Chamber Orchestra with Soloist

Repertoire intended for small orchestra and soloist can be difficult to record from the standpoint of balance and image. Quite often the solo instrument will be presented as "larger than life" in relation to the accompanying group, as it is exaggerated in size when placed directly under the main system. While this is not an issue when recording a large orchestra, the reduced number of players in a chamber group cannot compensate for this bias. Here the expanded tree works very well – as the main microphones are moved away from the soloist, the accompanying ensemble is then favored in the capture and not "dwarfed" by the lead instrument or voice (see Figure 17.2). This allows the spot microphones to bring the soloist's level up in the mix to the point where a suitable balance is achieved. Using a single main microphone in the center as opposed to a stereo pair in the middle will avoid having the image of the soloist pulled dramatically to one side or the other if they happen to be positioned slightly off-center. The soloist may also be placed outside the mains and looking back toward the group. This works especially well when the soloist is doubling as the conductor, as is normally the case in Vivaldi concertos.

17.4 Personalities of Chamber Orchestra Members

Chamber orchestras tend to be comprised of players with stronger personalities and opinions than regular orchestral musicians, as each member is required to take on more responsibility in a more exposed texture. A very good chamber orchestra with as few as three violinists to a part can be heard presenting Mozart or Beethoven symphonies with exceptional fullness and a well-blended string sound. Even very skilled string-section players in an orchestra may choose to "blend in" and simply follow their section leaders, but this is not an option for chamber music players. Each player has to produce a large output of sound with a precise focus on rhythm and intonation. Every member of the group is clearly heard at all times, so there is no hiding behind their stand partners.

It may come as a surprise to an outsider that many performance issues in chamber music are resolved by committee decision rather than by the conductor. In fact, many professional chamber orchestras have no conductor, leaving the driver's seat empty. The result may be an exceedingly democratic climate. In this case, the recording team should hope that most of these critical decisions have been made during rehearsals leading up to the actual recording session, so that discussions are kept to a minimum on recording time.

17.5 Chapter Summary

- Main microphones will be lower and closer than for orchestra capture.
- Balancing chamber orchestra with a soloist presents a new challenge.
- An expanded tree of three main microphones works very well for chamber orchestra.
- Be careful of strong personalities within the group – any member may act as the leader and should be taken seriously, so as not to insult anyone in the ensemble.

Chapter 18

Recording String Quartet

This can be one of the most politically charged recording experiences in an engineer's career. Sometimes the key is to perfectly match the group's natural timbre and balance with the most transparent microphone technique possible, although the acoustic balance as heard on stage by the artists may be interpreted differently once it is translated through the microphones and loudspeakers in the control room. Refining the microphone placement and balance or adjusting the stage positions of the musicians may be required to achieve the best results. Advice is offered in this chapter for best capture and for accommodating the personalities of the group.

18.1 Overview

Understanding the dynamics of a chamber group is critical when undertaking a recording session with a string quartet. That is, *dynamics* under the definition of "forces at work" or "underlying forces", as a political metric. The very design and structure of a string quartet requires that the group rehearses and performs together for years before achieving consistent results in balance and a unified and unique musical personality. Shaping the sound of the group is another huge investment of time and energy, and as such, the first and second violinists cannot simply exchange places. Playing second violin is a specialty and its role is to generate warmth while matching the phrasing of the first violinist. If the two violinists were to switch parts, the sound of the group might be completely different. Some of the most famous quartets in the world have histories spanning half a century, generally with only sporadic changes of personnel over the years, as each change in membership requires months and sometimes years of adjustment within the group.

Soloists such as pianists have enough trouble deciding how to interpret a piece they perform by themselves – imagine four musicians having to agree on how they play every note, phrase and theme. As these artists spend so many waking hours together, it is no wonder that a certain amount of tension might develop, and that the four personalities in the group may or may not be well matched as friends but may instead be more like colleagues with an exceptional gift for playing music together. Rumors abound of traveling quartets that ask to be booked into hotel rooms on different floors and at opposite ends of the building, so as to avoid seeing or hearing each other when they aren't rehearsing or performing together.

As discussed in Chapter 13, the politics of a recording session are usually more of an issue for the producer than the engineer. Everyone in the control room, however, needs to be aware of the social dynamic, and be careful not to openly question any strange comments or discussion between the artists. Observing and understanding a group's collective personality will help in deciphering their comments when working on the sound. It is critical that all four "voices" are evenly

DOI: 10.4324/9781003319429-23

represented, except in the case of deliberate balance changes within a piece that are specifically executed for musical reasons.

When it comes to recording string quartet there is an old saying – *"After hearing the first playback, if all the members of the quartet are equally unhappy, it can be confirmed that at least the balance is correct".*

18.2 String Quartet Seating

Several seating configurations exist for string quartet. Clockwise from front left, the most common is first violin, second violin, cello and viola, and the second most common is the same, except that viola and cello are reversed on the right side (see Figure 18.1). A less common approach is to split the first and second violinists left and right, with viola and cello in the back of the group (Figure 18.2), while an exceptional staging is for three of the members to stand while the cellist remains seated. One major issue in perspective is that the players in the back of the group tend to sound farther away than the "front row" of players, and this can be exaggerated if the placement of the main microphone system is not ideal. Positioning the microphones too far out in front of the group will exaggerate this disparity, especially when omnidirectional microphones are used.

The cello box: using risers for the rear players can help to compensate for their greater distance from the main microphones, just as the woodwinds may be placed on risers in an orchestral stage setup. One major issue with using risers is how the cello sound may be affected by them. Lifting the cello up off the stage physically decouples the instrument's endpin, which acts as a resonator,

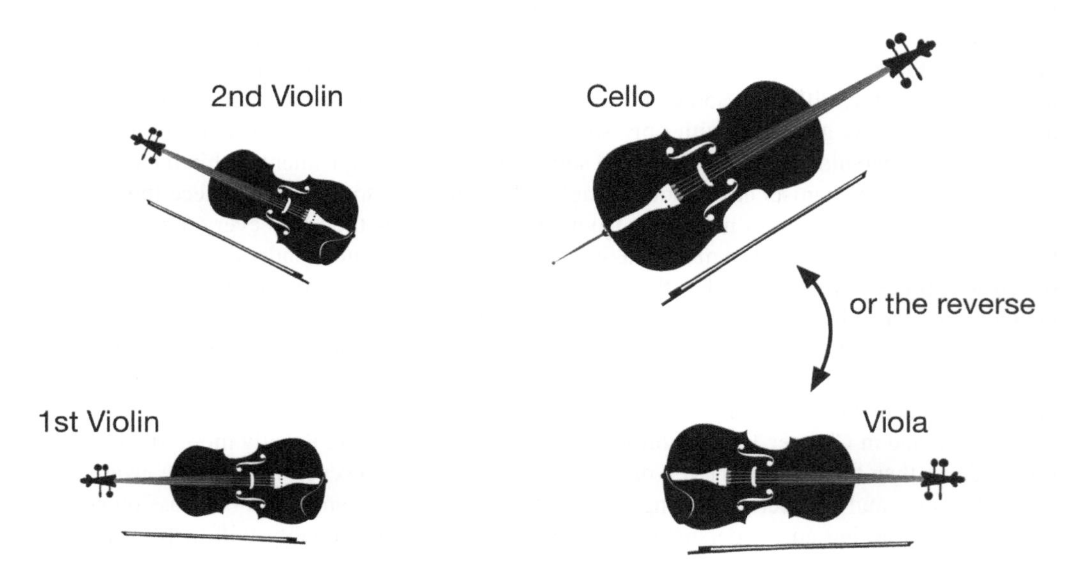

Figure 18.1 Diagram of the Most Common Seating Arrangements for String Quartet

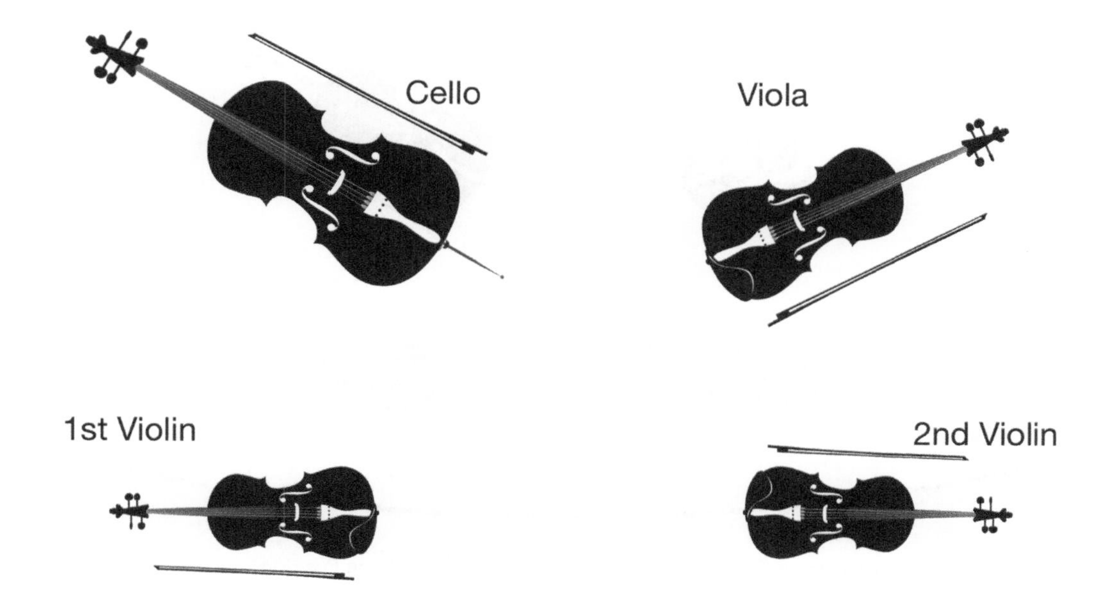

Figure 18.2 Diagram of a Seating Arrangement with the Violins Split Left and Right

from the floor. Just as a tuning fork can be amplified by placing the stem on any dense object such as a tabletop or the frame of a piano, the cello's endpin transmits energy into the floor, which helps support the low-frequency production of the instrument.

If a riser is to be used, it should be built in the style of a box with at least five sides, and possibly be open to the floor so that it resonates with the instrument rather than working against it and absorbing sound. Even a good cello riser can be hit and miss, depending on the exact place the endpin comes to rest (to some degree, this is the same for the stage floor). Experiments should be made with the cellist playing on and off the box, and in different positions, to determine the best placement of the endpin – this spot should be clearly marked with tape, so that each take is performed in the same position.

18.3 Recording the Quartet

Using two spaced microphones (AB technique) is one of the best ways to record string quartet, and using omnidirectional microphones will give the most honest and natural sound, with the richness of the lowest cello notes being properly captured. A Decca tree is technically a viable option for added control of the rear instruments and greater stability of the center image, although based on the author's personal experiences, this technique comes with its own practical concerns. The group may arrive at the session and notice that *three* microphones have been set up to capture *four* players – their first impression may be that one of the players will most likely sound as if they are "off-mic". String quartets are notorious for being sensitive to having individual members underrepresented in the balance (see previous caption box), and as such it is risky to give them any reason to support this tendency.

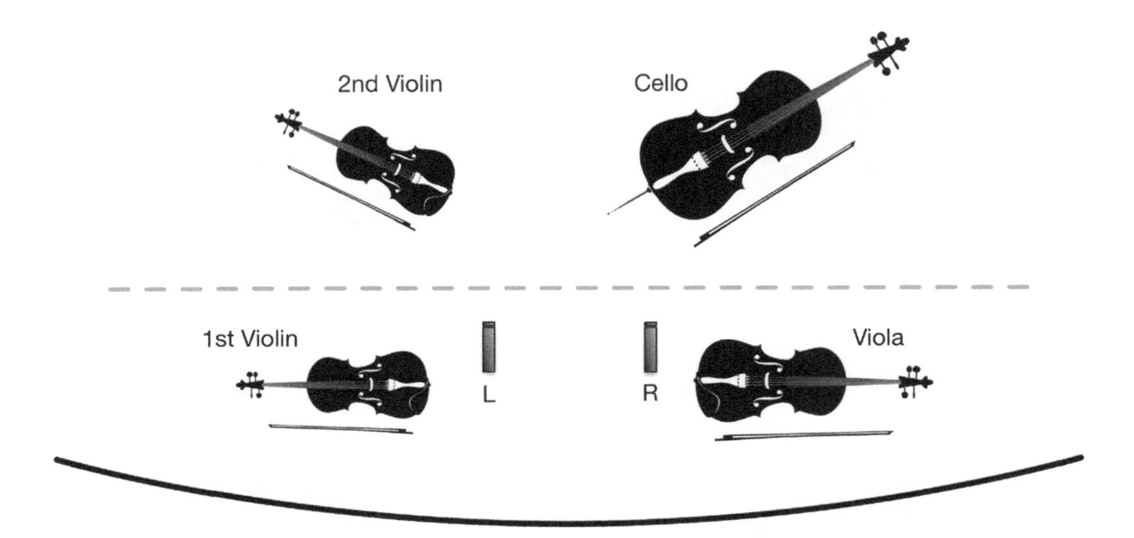

Figure 18.3 Diagram of a String Quartet on Stage with Starting Microphone Position

It is highly recommended to start with a stereo main pair, and if necessary, four spot microphones can be set up and positioned next to each member of the group, although they really should not be needed if the main pair is properly positioned with an appropriate height setting. In order to achieve an absolutely even balance of all four instruments, it will be necessary to get the main microphones close to a point directly above the line that virtually dissects the group front to back (see Figure 18.3). From that position, the microphones can be adjusted to optimize the balance based on the individual player's natural output. Moving the main system downstage a small amount from that point toward the front row will add a small amount of depth to the ensemble's perspective without upsetting the overall balance. The microphones can then be adjusted in angle so that they are looking directly at that invisible line between the two rows, or even farther back into the second row if necessary, based on some careful listening to the placement.

EXERCISE 18.1 RECORDING STRING QUARTET

1. Using a single tall stand with a sturdy boom and a wide bar, set up a pair of omnidirectional microphones at a spacing of around 70 cm (28 in.) and place them up and over the quartet, facing down, at about 3 m or 10 ft. This is also possible with two separate stands, but the extra legs will decrease the amount of open real estate around the musicians' feet.
2. Locate the exact middle of the group from front to back and place the pair ever so slightly closer to the front row of players, while adjusting the angle of the microphones

so that the rear players are as much or slightly more on axis than the front row. This should give the right amount of depth in the recording without causing a disparity of perspective or presence between the front and back players.

3. Adjust the height to obtain the desired balance of direct-to-reverberant sound.

Marking chair positions: when recording with only two microphones directly over the group, the physical position of each quartet member is crucial. Every time the quartet takes a break or heads to the control room for a playback, the chairs may move and will need to be returned to their original positions before more recording takes place. As such, the engineer needs to follow the group back to the stage or into the studio to check this very important detail throughout the recording process.

It goes without saying that the quartet should be included in some or all of the playbacks while this very specific position for the main system is being refined. Any other inconsistencies can normally be adjusted by changing an individual player's chair position. For instance, if the viola is too loud it might help to move the violist's chair slightly away from the main system to correct the overall balance. The more the quartet is involved in the steps being made to improve the sound, the easier it will be to work productively with the group throughout the session. In this fashion, the musicians will feel as if they helped to mold the *recorded* sound, not just the sound produced acoustically on stage, leading to a stronger collaboration between the quartet and recording team.

See the companion website for string quartet examples (https://routledgetextbooks.com/textbooks/9781138854543):

1. Audio comparison of main microphones out in front of the group compared to main microphones up and over the quartet, and
2. Video of the quartet showing the main system in the latter (preferred) position.

NOTE: The spot microphones shown in the video are only meant to demonstrate possible placement. They were not used in creating the audio examples or the soundtrack of the video.

18.4 Chapter Summary

- Be aware of personality conflicts within the group, and how these might affect the balance.
- Two main microphones over the group is the recommended approach.
- Involving the group in the process of refining the sound and balance will certainly help relations and gain the trust of the musicians.

Chapter 19

Recording Woodwind and Brass Quintet

These two chamber music groups are rather similar in terms of stage setup and the necessary approach to recording. Starting microphone positions can be compared to the suggested string quartet setup: refer to Figs. A.4, A.5 and A.6 in the Quick Start Guides (Appendix).

19.1 Woodwind Quintet

The woodwind quintet is a chamber music group made up of five completely different instruments, each having their own specific timbre and diffraction pattern. Flute, clarinet, oboe and bassoon are joined by French horn to make up this very unique ensemble. While the first quintets were written somewhat earlier, the genre was really established in the early 1800s. A large repertoire exists for this group, with the most prolific output coming from the 20th century.

19.2 Woodwind Quintet Seating

Standard seating for woodwind quintet is flute and oboe left to right facing each other in the front row, clarinet and bassoon left to right in the second row, and French horn facing forward in the rear center of the group. This forms a rough semicircle, in which sightlines are clear and musicians can easily hear one another (Figure 19.1).

The best approach for an even capture is to focus on the main pickup, using roughly the same approach as with the string quartet. The initial placement will be slightly outside of the group toward the audience, but not by much. The same acoustic modifications should be made after listening with the musicians, to help balance the ensemble on stage in a natural way. Adjustments to the balance can be made by shifting chair positions closer to or farther from the main microphones, and the musicians will most likely adjust their performance after hearing a playback of the captured sound. Spot microphones can be used as a last resort, but they should not be relied upon to "build" the sound or create the basic sound of each instrument in the overall mix. One point to keep in mind is that these five instruments are quite different from each other, and each should portray their own unique sound, even though an even blend is desired. French horn will sound more diffuse, as its bell faces away from the ensemble and the microphones, but this is of course the natural sound of the instrument. A small amount of signal from a microphone placed behind the horn may help to add a little focus to the overall mix.

19.3 Brass Quintet

Also dating from the 1800s, the brass quintet as an ensemble has a more homogenous timbre than a woodwind quintet. The modern brass quintet as we know it today was only formed around 1950.

DOI: 10.4324/9781003319429-24

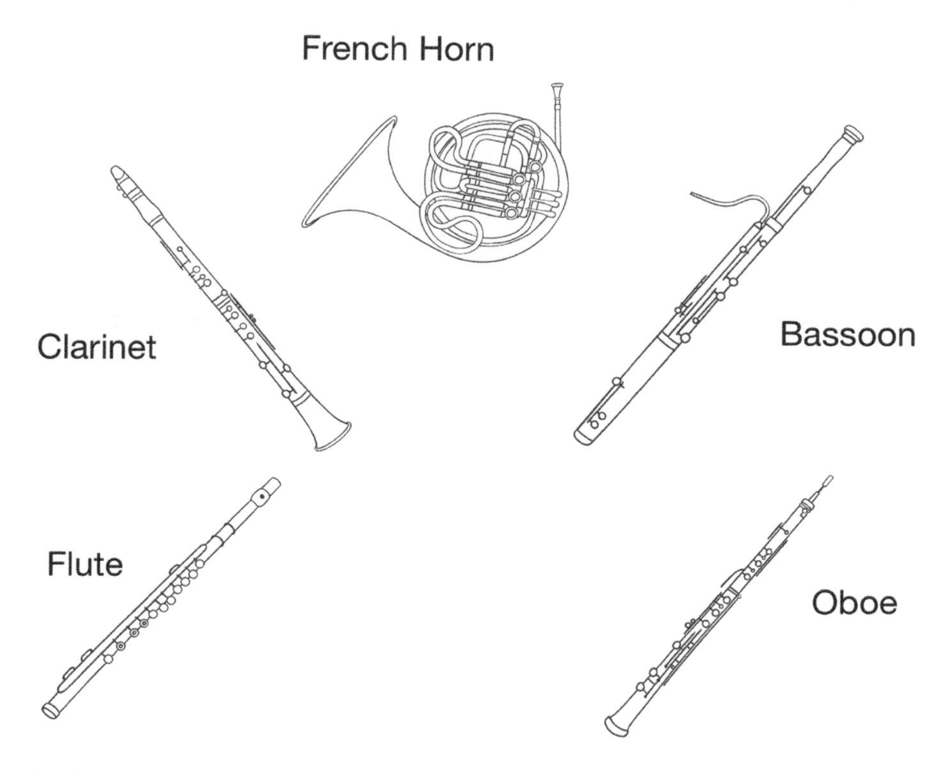

Figure 19.1 Diagram of a Typical Woodwind Quintet Seating

While the quintet is the most common brass chamber group, other brass ensembles of different size and makeup are often formed to perform various repertoire.

Although the ensemble is typically made up of two trumpets, trombone, French horn and tuba, the trombone is occasionally swapped for baritone or euphonium and the tuba is sometimes replaced with bass trombone. Trumpetists may occasionally switch to cornet or flugelhorn, depending on the repertoire. Sound propagation is quite similar to that of the British-style brass band. The ensemble tends to direct sound into the hall at a low altitude, so typically, a higher microphone setup will miss a great deal of the clarity and brilliance of the trumpets. For this reason, the starting main microphone position should be farther back from the group and at a lower height than string quartet, although this does not favor the French horn and tuba. A careful compromise needs to be made between trumpet clarity and horn presence.

19.4 Brass Quintet Seating

Typical seating normally has the two trumpets in the front row facing each other, followed by French horn and trombone across from each other and the tuba or bass trombone in the rear center of the group. This is the most common seating, although some groups sit differently and sometimes prefer to stand while playing. The result tends to resemble a "box" or "U"-shape formation. In this case and whenever possible, the ensemble should be encouraged to sit in more of a "V" shape so that they project more directly into the hall and into the main microphone system. As with string

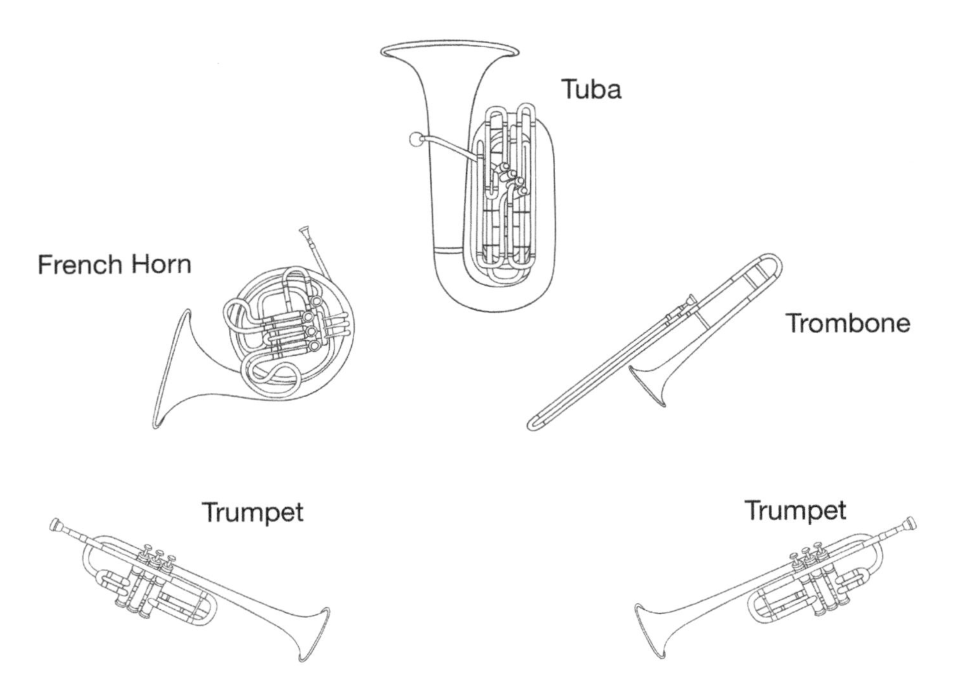

Figure 19.2 Diagram of the Most Common Brass Quintet Seating

quartet and woodwind quintet, adjustments to the balance can be done by moving chairs closer to and farther from the main microphones, and the musicians will adjust their performance based on hearing a playback of a test recording excerpt. Figure 19.2

19.5 Chapter Summary

- Woodwind and brass quintets can be approached in a similar manner to the string quartet, whereby the overall balance is captured on the main system with physical adjustments made on stage (chair positions, etc.).
- A woodwind quintet is an ensemble of instruments with differing diffraction patterns, although this anomaly helps distinguish each voice.
- Brass quintet instruments (trumpet, trombone) are more directional in nature than woodwinds and this should be taken into account; a brass quintet usually requires a lower height setting than a string quartet or woodwind quintet.

Chapter 20

Recording Piano with Other Instruments

Balancing the weight and power of a 9-ft. concert grand piano with a small but determined violin is a huge challenge. The piano invariably excites the room much more than the violin, and it masks the natural reverberation of the violin without any effort. This chapter offers solutions for balancing piano with various instruments, including voice.

20.1 Piano as a Partner in Chamber Music

The biggest discrepancy in sound between the piano and any other instrument in a chamber music recording is the amount of "room effect" caused by each instrument in the group. Creating a homogenous presentation of all the instruments in the ensemble is the biggest challenge, followed by balance. The greater the influence of the recording venue on each source, the more that source will sound recessed in the overall picture. A piano will excite the room or hall much more easily than any other instrument, causing an imbalance of perspective.

To make matters worse, the piano is traditionally positioned farther back on stage behind the other instruments, and for good reason – so it doesn't block the audience view of the other musicians, and because it is much louder. A less positive outcome is that the piano sounds recessed as compared to the instruments in front. This is less of a factor for the audience, especially those who are seated at a greater distance from the stage, but for microphones placed on stage during a recording session, it is a major concern.

20.2 The Piano Lid

Using the short stick or completely closing the lid of a piano will not make it sound much softer. The piano will be almost as loud, but with a dull, thick sound that will work against the recording team as they try to enhance the clarity of the instrument. If and when the lid is visually in the way the recording team might consider removing it completely, but even this dramatically changes the sound and directivity of the instrument and will further reduce the focus of the sound into the hall and main system. This is more likely to be an issue in orchestral or large ensemble recording than in chamber music. In live concert recording, the audience may notice a lack of piano into the hall, while the resulting recording will be fine. The engineer can mostly compensate for the lack of focus and directivity the lid normally provides by placing spot microphones directly over the soundboard, as opposed to outside the frame of the instrument.

DOI: 10.4324/9781003319429-25

Concert grand pianos are designed to be played with the lid fully open, resulting in a balanced sound from low to high frequencies. The lid is basically a reflector of middle and high frequencies and should not be thought of as a "volume control".

20.3 Piano and Violin

In concert, violinists prefer to stand to the left of the piano keyboard (as viewed from the audience), for best communication with the pianist. This is, of course, less than ideal for recording, as the violin is localized on the left and the piano sound will be coming mostly from the right side, unless a radical approach is taken regarding the placement of the main system (see Figure 20.1). Trying to convince a violinist to stand in the "crook" of the piano (where the frame becomes narrower halfway down the instrument), so that the violin will be centered in the recording, may be a losing battle. Firstly, they will not be used to hearing the piano in this way, and secondly, they may have difficulty hearing themselves, especially during more intense or rhythmic passages of music. Having the violinist in this position is the best option for concert recording, but it limits possibilities for piano spot placement and affords virtually no isolation of the piano in the violin pickup.

Violin facing piano: for recording sessions, the best option may be to have the violinist (or other instrument) facing directly toward the piano and to place the main microphone system between the two instruments, as in Figure 20.2. The main pickup can be adjusted between the two instruments to achieve the right balance and perspective, and the overall direct-to-reverberant ratio can be addressed by adjusting the height. Since most omnidirectional microphones approach a cardioid

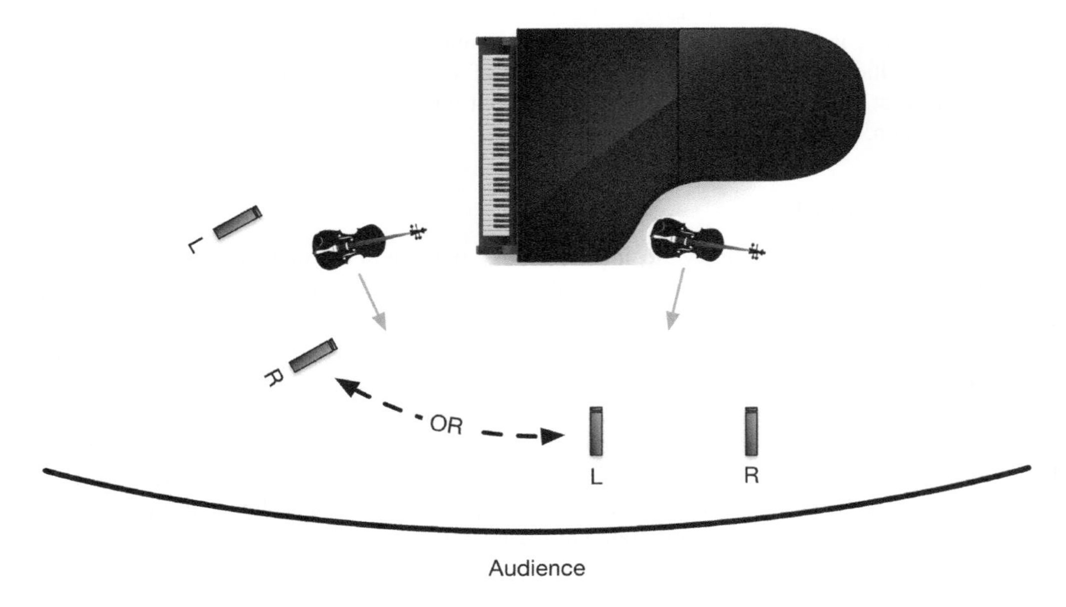

Figure 20.1 Diagram of Two Typical Concert Positions for Violin and Piano, and the Respective Placements of the Main Microphones

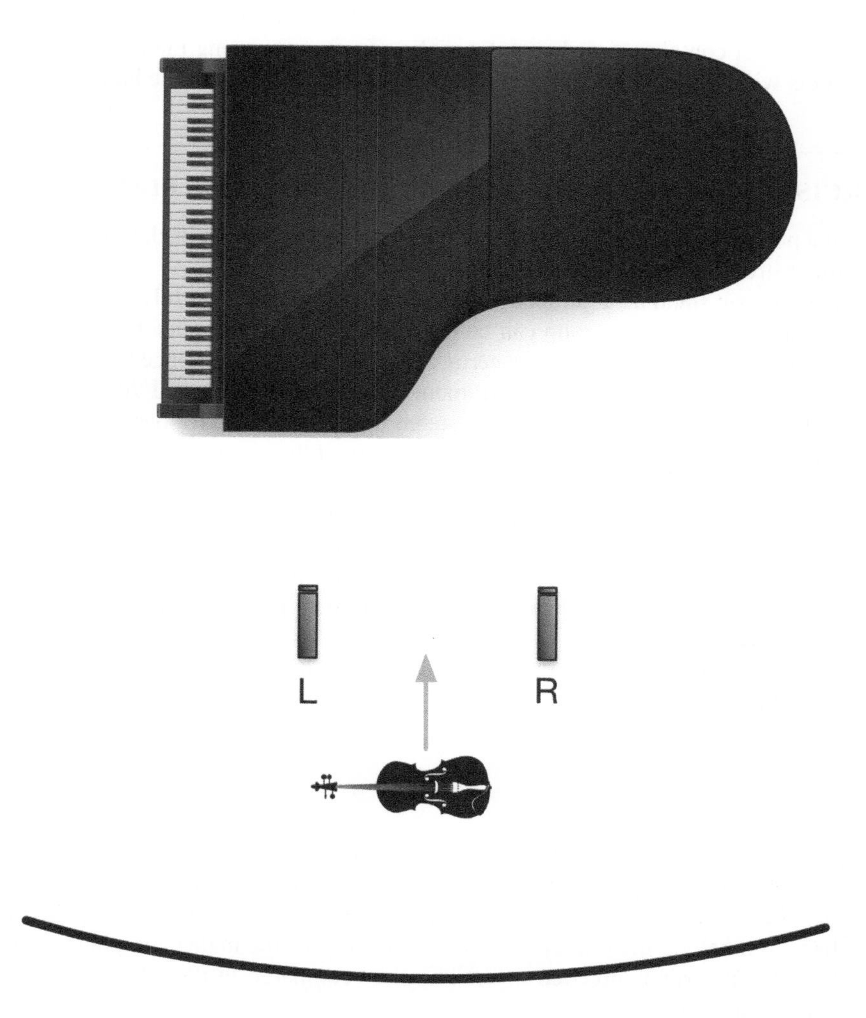

Figure 20.2 Diagram of Recording Setup with Violinist Facing the Piano, and the Starting Position for the Main Microphones

polar pattern in very high frequencies, the microphones might be oriented directly toward the floor for an even capture of both instruments. Alternatively, a decision could be made to favor the high-frequency content of one instrument over the other, depending on the situation.

Even with a slight increase in distance between the two players, the improved line of sight makes for easy visual communication. Violinists can hear themselves very well in this position, and within a short amount of rehearsal time, they normally adjust quite easily to hearing the piano from directly in front rather than their left side. For the spot microphones, assuming cardioid polar patterns, the leakage of each opposing instrument is greatly reduced, allowing for better control over balance in the mix. In other words, the violin spots are facing away from the piano and only the violin is on axis. The same is true for the piano spots; however, this is less significant, as it is usually the violin that needs more help to achieve a suitable balance in the mix.

See the audio example on the companion website (https://routledgetextbooks.com/textbooks/9781138854543); it contains recordings of a violin sonata, comparing main pickup, violin spots and piano spots with violin facing the piano vs. violin facing out toward the hall.

EXERCISE 20.1 RECORDING A VIOLINIST FACING THE PIANO

1. Plan a position for the violinist out in front of the piano looking back toward the middle of the instrument, as in Figure 20.2 above.
2. Using a music stand as a guide, set the violin position to be slightly more distant than necessary. This way the group can be brought closer together once they have seen the setup and have expressed their concerns over the unfamiliar distance between them.
3. Set the mains between the two instruments at a spacing of around 70 cm or 28 in. to start. Use separate stands rather than a stereo bar on one stand for best flexibility and to ensure that the violinist has an unobstructed view of the pianist.
4. Bias the placement of the main system toward the violin to help in balancing the two instruments. Regarding perspective, the piano is already much closer to the main system than if the violinist were placed in front of the piano looking out toward the hall.
5. Adjust the left and right main microphone so that the piano is balanced evenly across the left and right channels, and then "center" the violin by physically adjusting the position of the musician.
6. Starting at around 2.75–3 m or 9–10 ft., adjust the height of the mains to achieve the desired direct to diffuse ratio.

When positioning musicians on either side of a main system, be sure to maintain "audience perspective" for all stereo pairs of microphones. In other words, the violin left and right spots should be labeled and connected as seen from that player's perspective, facing the capsules, rather than from behind the microphones as is the case for the mains and piano spots. This will ensure that no pair is inverted in the image as compared to the other pairs of microphones (see Figure 20.3).

20.4 Piano and Cello

Because the cello is held and played at a rather low height, it is possible to achieve a good balance with the cellist sitting near the crook of the piano and facing out, as is typical for singers in concert position (and shown in the right-hand side of Figure 20.2). With the cello situated in a lower position than violin, the spot microphones will be somewhat protected from the piano sound, as that energy is mostly directed out and up toward the main system. In this configuration, it can be difficult to place the piano spots in an ideal position, but minor compromises can be made. Having the cellist face toward the piano in the context of a recording session is also viable, for all the same reasons laid out in Section 20.3. Improved visual communication is the strongest argument, followed by control over perspective and spot microphone leakage.

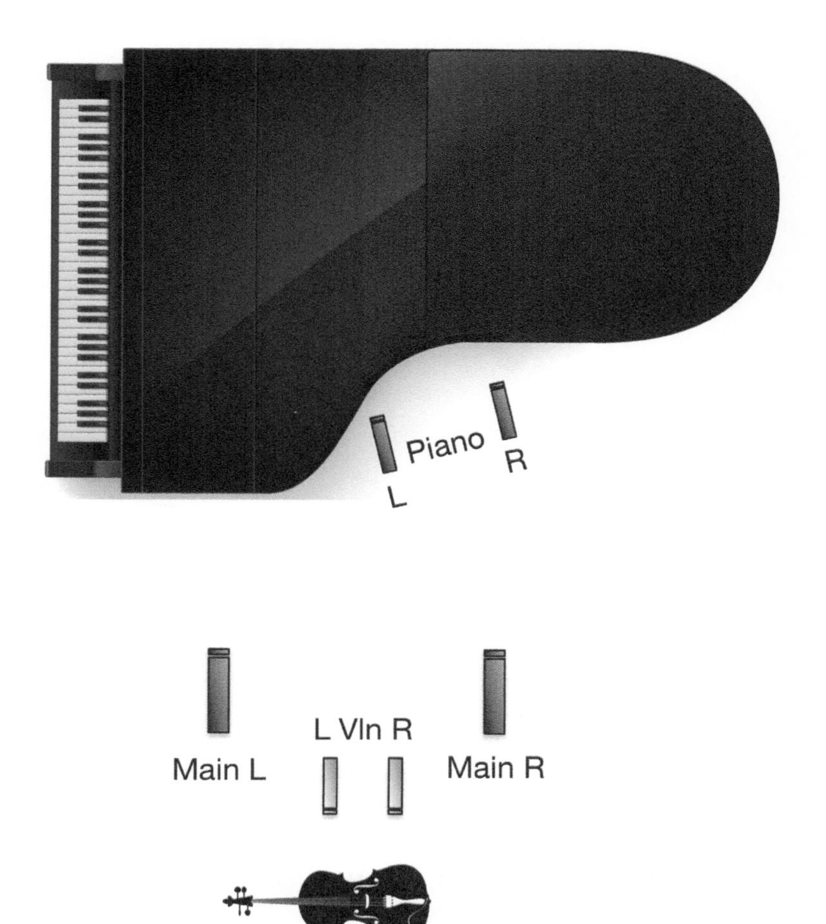

Figure 20.3 Diagram of Recording Setup with Violinist Facing the Piano, and the Various L/R Pairs Conforming to "Audience Perspective"

Placing the cello on a box or riser helps to raise the instrument toward the main system, but this may work against a unified overall presentation, especially if the piano is already more recessed than the cello and requires improved focus. Again, the box should be "auditioned" by having the cellist play on the box and then on the floor. If the cellist is placed so they are facing the piano, the main system can simply be moved toward the cello to obtain the desired perspective, especially if the cello sound is more suitable without a riser at all. If the cello sound is deemed superior because of the box, adjustments can be made in reverse by moving the main system upwards or toward the piano to compensate for the box height.

20.5 Piano Trio or Quartet

As instruments such as violin or cello are favored in the main pickup in an effort to balance the group, they end up sounding much more direct and drier than the piano, as they are closer to the

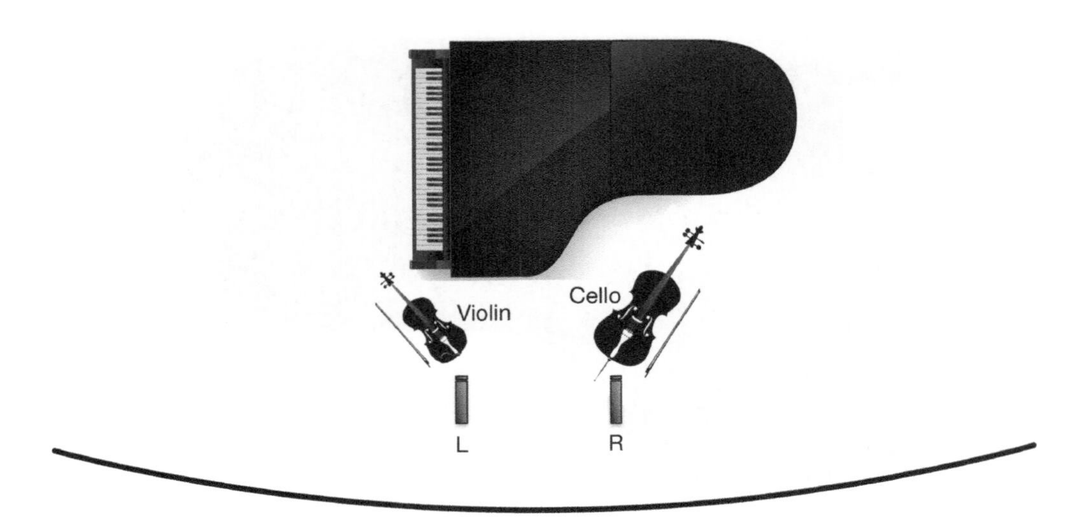

Figure 20.4 Diagram of Recording Setup for Piano Trio

main microphones (see Figure 20.4). In the traditional seating of a piano trio, comprised of violin, cello, and piano, the microphones can be brought up and over the strings so that the main system is less distant from the piano while at the same time the string instruments remain suitably present in the overall capture. Tucking the piano in as tightly as possible to the group can help to even out the disparity, as long as all of the musicians are comfortable with the close proximity.

EXERCISE 20.2 ADDING FOCUS TO THE PIANO IN A CHAMBER GROUP RECORDING

1. Once the main system has been optimized and the depth of the group has been minimized, the addition of two cardioid piano spot microphones will help to bring the piano the rest of the way into focus.
2. A narrow spacing is best – around 50 cm or 20 in. – since the piano image will be quite wide as heard in the main pickup.
3. The microphones will need to be fairly close to the piano, but not quite inside the instrument. This will ensure that the capture is mostly piano signal, and not string leakage.
4. Open up the piano spots gradually into the mix with the main system, just to the point where they "dry up the piano" or reduce the room effect without making the instrument much louder in level.
5. Mute and unmute the piano spots to evaluate the difference in perspectives.

For recording sessions, the strings might be placed out in front of and facing back toward the piano, with the main system in between piano and strings. There is less need, however, for groups such as a trio or quartet to be taken out of their traditional and rather familiar seating arrangement for the sake of the microphones.

20.6 Piano and Voice

Singers will benefit greatly from facing the piano, mainly because they can hear themselves more clearly. The recording engineer will find it easier to capture the piano and voice with an adequate amount of separation in the spot microphones. In some halls, however, there will be a distinct lack of vocal projection into the room and the natural reverb will be dominated by piano sound. Placing the vocal spots so that the piano is facing the rear of those microphones will greatly reduce leakage and allow for improved control in balancing the vocal into the overall mix. As presented in Section 12.2, be careful to check how the vocalist's music stand is placed, and that the singer has not created a direct reflection of their vocal sound into the spot microphones or the main system.

There may be times when a singer will ask for the lid of the piano to be placed on the short stick, mostly likely so that they can hear their own performance more easily. Singers will be able to hear themselves better when facing the piano, and an accomplished pianist will be able to adjust their dynamics so that the balance is satisfactory for everyone with the lid fully open. A vocal recital album with the dull, thick sound of a closed piano will be a real disappointment and should be avoided at all costs.

20.7 Chapter Summary

- Piano excites the room much more than other instruments and will need to be controlled.
- Many instruments benefit from facing the piano as opposed to facing out into the hall.
- Placing the main pickup between the two musicians allows for a more direct piano pickup and less leakage in the spot microphones.

Chapter 21

Recording Solo Piano

There are many available resources for information on recording piano and many proposed techniques. Hundreds of recordings are available as references spanning the mainstream repertoire, but it is virtually impossible to find out how each recording was captured (number of microphones, microphone type, etc.). For the newcomer, this can be a frustrating and discouraging experience. This chapter provides a summary of the most common techniques for classical presentation of the instrument and some suggested starting points for evaluating piano sound over microphones.

Audio examples for this chapter are available on the companion website (https://routledgetextbooks.com/textbooks/9781138854543):

1. Solo piano recording comparing the techniques AB, ORTF, XY, M-S and Blumlein.
2. Video showing all the microphone placements in relation to the position of the piano.

21.1 The Recorded Sound of the Piano

It may seem elementary, but the recorded sound of a piano should actually sound like … a piano. That having been said, the question arises – what are the parameters that constitute an exceptional piano recording? There are many different opinions on exactly what these might be. To revisit the analogy from Chapter 1 – how much salt is too much salt? – the amount of preferred brightness or bass frequency that is considered to be appropriate will vary greatly in solo piano recording – much more than with orchestral repertoire. Image is another heavily debated issue. Some practitioners enjoy filling the entire space between the loudspeakers with piano sound, whereas others prefer a more natural (or narrower) presentation of the piano image residing inside the loudspeakers, while the reverberant sound of the room fills the extreme left and right sides of the lateral soundstage.

Because of all these differences, it is a good idea to head to a music library or the Internet to compare a few versions of existing recordings of the repertoire that is to be recorded. If possible, four or five different recordings should be evaluated to get a good overall sense of what is considered to be acceptable and to decide on a personal preference. Again, the artist and/or client will need to approve the final sound, but the engineer needs to find a starting balance, based on their own aesthetic, that is then presented to the artist.

Many excellent pianos exist on the marketplace, although several companies have monopolized the market for many years. Having sold over 600,000 pianos to date, Steinway and Sons has been the clear frontrunner of all manufacturers, with pianos made first in New York,

DOI: 10.4324/9781003319429-26

beginning in 1853, and afterwards in Hamburg. American competitor Baldwin grew from the late 1800s, and now builds its pianos in China. Austrian company Bösendorfer – now a subsidiary of Japanese instrument maker Yamaha, which has its own line of fine concert grand pianos – and the relative newcomer Fazioli, from Italy, are some other examples of the many familiar names of excellent piano manufacturers worldwide.

21.2 Various Techniques for Recording Piano

It is amazing how many possible recording techniques have appeared in print regarding piano capture for both pop music and classical applications. Suggestions range from taping boundary layer microphones inside the lid to lowering capsules into holes in the soundboard. Recording piano for classical repertoire requires that the piano be presented in a manner that offers the most natural possible result, with stable imaging and even coverage of the instrument.

Traditionally, piano recordings are presented from the audience's perspective rather than that of the pianist. With microphones placed out in front of the instrument, the higher range of the instrument should appear toward the left side of the image and the very lowest notes will be heard on the right, although the main body of sound should be localized evenly across the middle of the soundstage. There is virtually no separate localization of the pianist's hands, as the left and right microphones work together to capture the entire instrument under one system. This differs greatly from what might be possible in a very tight and widely spaced "pop music" technique, designed to capture the low and high range, or left and right hands, with maximum separation.

One technique that has been common in Europe for many years is to place a pair of microphones in a narrow spacing (around 30–40 cm or 12–16 in.) positioned outside the piano, at 90° from the body and quite near the hammers, as shown in Figure 21.1. This is frequently combined with a wider pair placed at a greater distance from the piano, so that the amount of "room sound" can be adjusted in the control room by changing the level of the second pair of microphones in the balance. This is a similar approach to the "small ab/ big AB" technique for recording orchestra as presented in Chapter 6. Capturing the direct sound of the piano near the hammers will provide enhanced clarity and brilliance, although at times the result may be overly aggressive, especially when a very bright instrument is in use.

The British company Decca Records used a very specific technique quite consistently over many decades, and this technique is still common. Its preference was for two omnidirectional microphones placed in parallel and at a narrow spacing toward the tail of the instrument, also shown in Figure 21.1. In this position it is possible to satisfactorily capture low frequencies with an appropriate amount of brilliance. The distance from the piano can be adjusted to taste for the balance of direct to diffuse sound. More detail and background on this technique is available in [1].

What is the engineer's most useful tool when it comes to evaluating the sound of a piano? The ear. Listening while standing in front of a piano will help in locating the ideal placement of the microphones, as every piano will have a slightly different timbral characteristic, and every hall will have a different influence on the instrument.

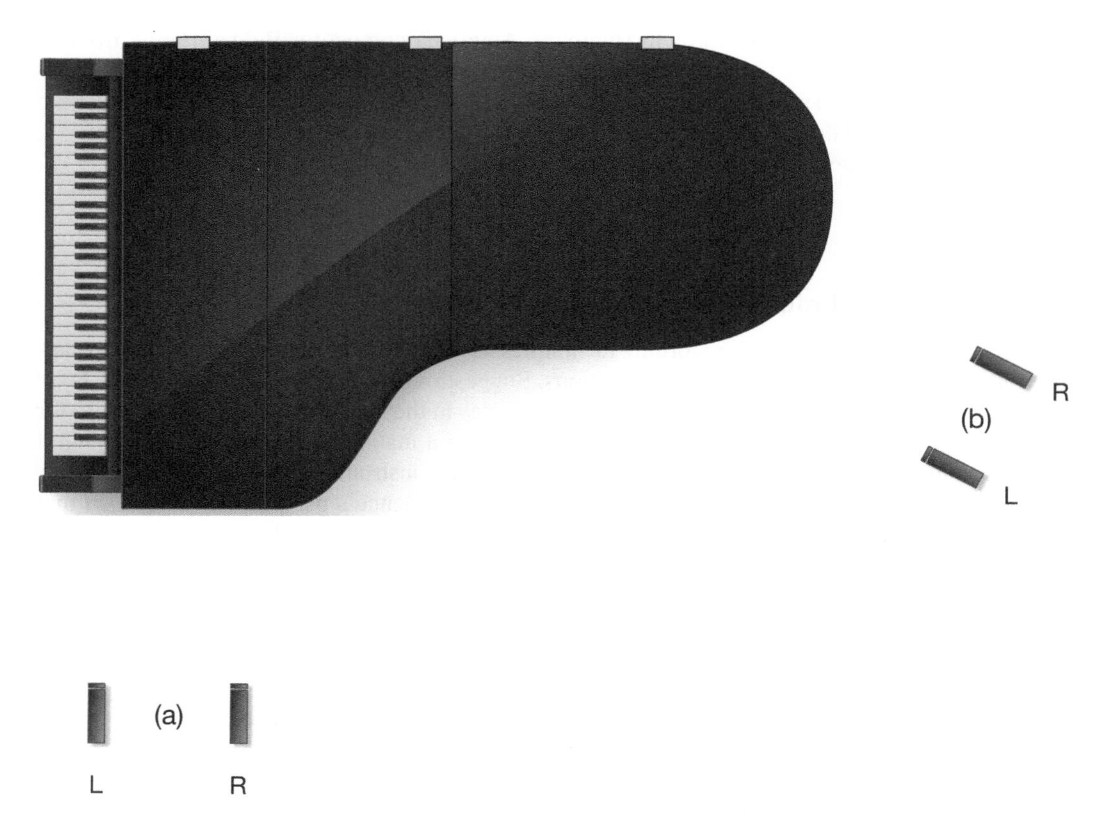

Figure 21.1 Diagram of Piano Showing Two Techniques: (a) Microphones Placed Near the Hammers; (b) Microphones Placed Near the Tail of the Piano

Engineers new to classical music recording should take the time to experiment with all available techniques and are encouraged to discover for themselves their particular preference. However, a compromise is suggested here, with a placement somewhere between the hammers and the tail. On most instruments, this is a position where the direct sound tends to be less aggressive than nearer the hammers, and the midrange is fuller and more present than the tail area. Although a relatively wide spacing will result in an image with slightly reduced stability, the instrument will be presented with substantial impact and even energy throughout the range of the instrument (see Figure 21.2).

Only two microphones are needed for this technique, and their distance to the instrument can be adjusted to obtain the desired balance of direct and reverberant sound in the room. Adjusting the point of capture between the hammers and the tail will dramatically change the timbral characteristic of the sound, just as varying the microphone height will change how much reflected sound from the lid is included in the capture along with the direct sound of the instrument. Changing the spacing of the microphone pair will vary the width of the source image.

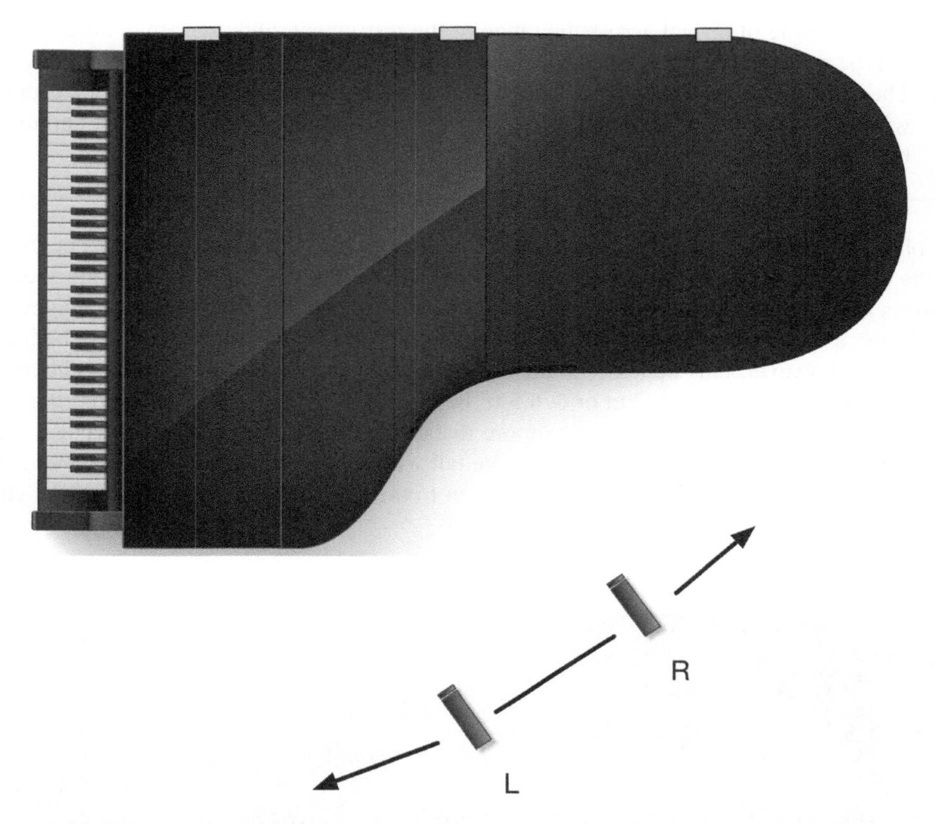

Figure 21.2 Diagram of Piano Showing Microphones Placed Midway between the Hammers and the Tail

EXERCISE 21.1 RECORDING SOLO PIANO WITH TWO MICROPHONES

1. Position the piano close to the edge of the stage. Have the artist play and listen to the instrument in the venue on stage, near the position the microphones will be placed. This is a good opportunity to listen for any mechanical noises that the piano technician may be able to solve, including a noisy bench.
2. Try moving the piano around, for instance slightly upstage (away from the audience seats) or downstage, if the sound is lacking in some regard (brilliance, low-frequency support, reverberation).
3. Once the physical placement seems suitable, place two omnidirectional microphones in front of the piano at a spacing of around 80–90 cm, or 32–36 in. The distance from the microphones to the body of the piano might be 1 m (3 ft.) or more and will be adjusted based on the amount of reverberation in the hall.
4. Using the hinge of the lid closest to the lowest note on the piano keyboard as a focal point, adjust the two microphones so that they are at equal distance from that point (see Figure 21.3). This should provide a fairly even balance of left and right channels in

terms of time of arrival across the range of the instrument, although small adjustments will be needed after listening in the control room.

5. The microphone placement can then be shifted either toward the hammers, for a capture with enhanced clarity and brightness, or toward the tail, for a richer, warmer and more blended timbre.

Microphone height is set "to taste", although it is recommended that some reflections be included from the underside of lid, as this reflected sound will add an impression of power to the sound and help focus mid and high frequencies. Positioning the microphones slightly higher up may result in more clarity, as most of the reflected sound from the lid passes underneath the microphones. The overall result, however, may be less vibrant and complex in timbre, and therefore less interesting. A starting height of around 1.8–2 m or 6–6.5 ft. is suggested, depending on how close the microphones are placed to the instrument, as a more distant placement will typically dictate a greater height setting. Engineers who happen to be quite tall will be able to listen to the instrument at this height, otherwise a step stool can be used to listen to the piano at the same height as the chosen microphone position.

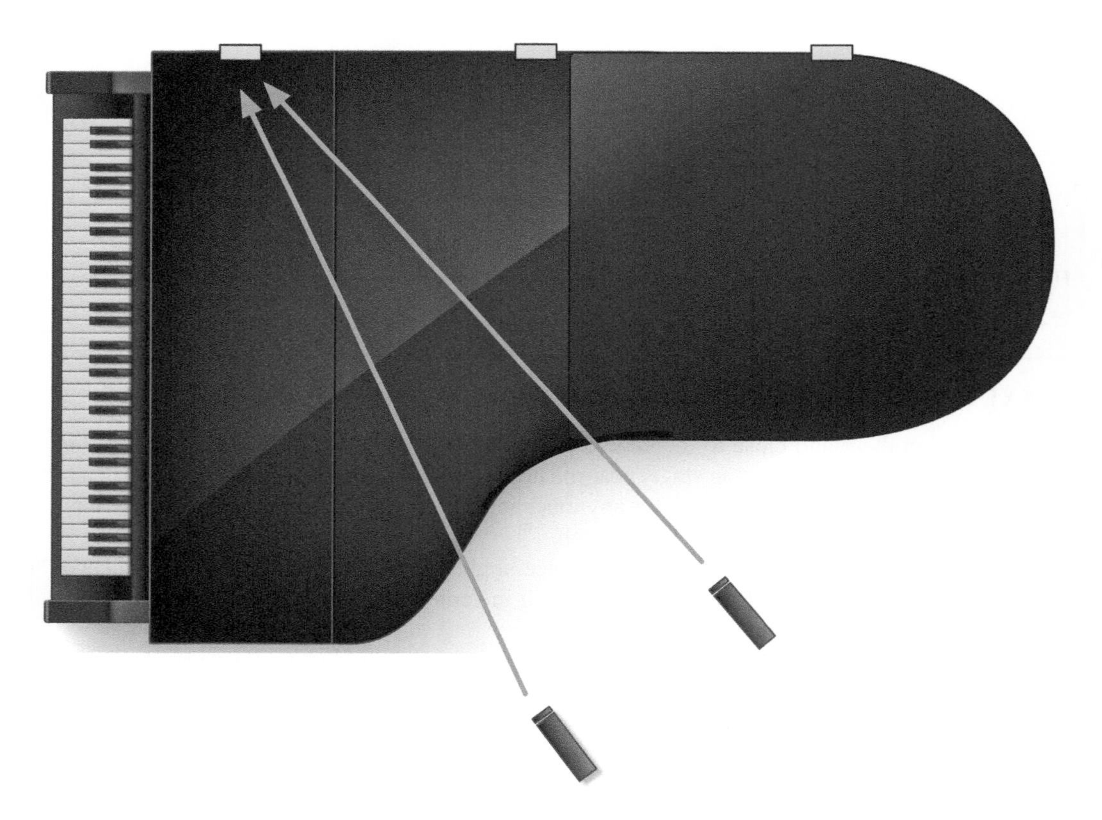

Figure 21.3 Diagram of Piano Showing Microphones Equidistant to the Hinge of the Lid That Is Closest to the Keyboard

With the rather wide spacing that is proposed here, it is worth listening in front of each single microphone, to locate the most "interesting" placement for each capsule. In other words, if the engineer is searching for a more brilliant sound, a separate position for each microphone can be found that is rich in high-frequency content. A search for each single microphone placement may be performed, as long as the engineer bears in mind the concept of a spaced pair as a stereo system. Needless to say, there will be many trips back and forth from the stage to the control room. This can normally take place while the pianist is warming up, so as not to waste too much recording time during the session. As with all classical music recordings, audience perspective should be adhered to, with the higher notes of the piano sounding "leftish", and the most extreme bass notes coming from the right-hand side, along with balanced sound from left to right.

Once the artist has heard and approved the sound after a few refinements, the position of the piano and the microphones should be marked with tape. On a break, or before leaving at the end of the first day, the positions should be measured and documented, in case anything "moves" overnight (due to cleaning staff, piano technicians or other users of the hall). Placing tape on the floor is quick and convenient, but it is nowhere near permanent.

21.3 Two Pianos, Four Hands

"Concert position" for pieces with two pianos usually requires one piano to be placed in front of the other with its lid removed. In this case the single lid of the rear piano is used to project the sound of both pianos into the hall (see Figure 21.4). When recording two pianos together, it is hard to balance the instruments, as there is little separation between the two "interlocked" instruments and the pianos invariably sound different since one instrument is without its lid.

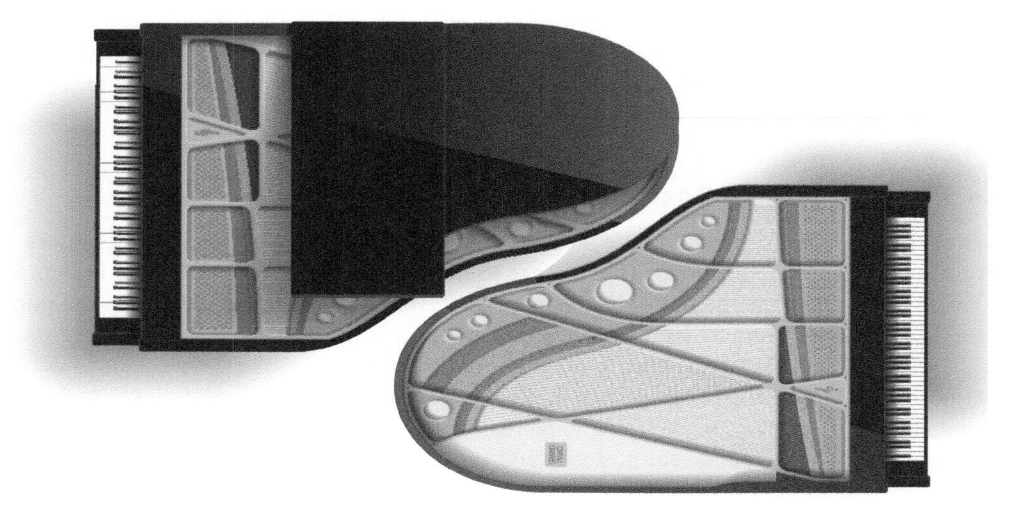

Figure 21.4 Diagram of a Typical Two-Piano Placement for Live Concert Events

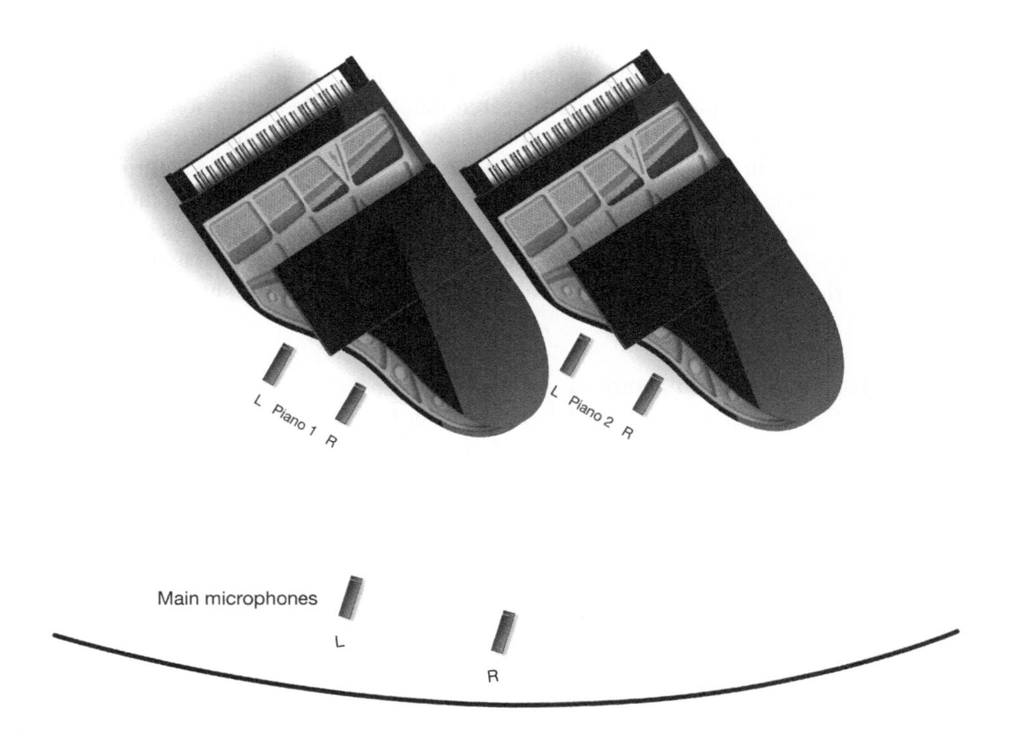

Figure 21.5 Diagram of a Possible Two-Piano Placement and Microphone Positions for Recording Sessions

A better solution is to place the two pianos in parallel, with both tails facing out to the house at a diagonal to the centerline of the hall, as in Figure 21.5. This way, both instruments can be recorded with their lids on and fully open. One pair of main microphones can be used to capture the two instruments together. Spot microphones added to each piano help in controlling the balance of the "two voices" as needed and the lids of each instrument help to reduce leakage between the two sets of spot microphones.

21.4 Piano Recording Documentation

A simple method for documenting the distance from the microphones to the piano is to use the outside two hinges of the piano lid. The distance from each microphone to each hinge is recorded, so that an exact recall can be performed as necessary. Heights and distance between the microphones should also be noted (see Figure 21.6). This sheet is also available for download on the book's companion website (https://routledgetextbooks.com/textbooks/9781138854543). The exact position of the piano on stage should also be documented on the same sheet, and photos will help to jog the memory when trying to recreate a setup. It may be interesting to keep track of the serial number of the piano, for posterity and for future reference. Rented pianos can change dramatically depending on the demands of the pianist, so there is no guarantee of consistency through requesting the same piano over consecutive sessions, although it is still worth keeping track of such things.

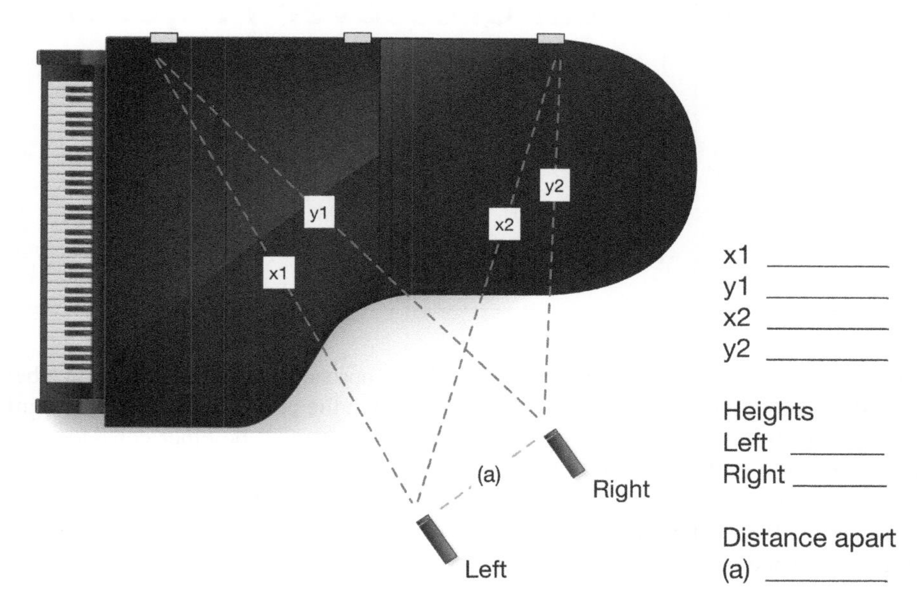

Figure 21.6 Chart for Documenting Measurements from Each Microphone to Each of Two Hinges, Along with Microphone Heights and Spacing

21.5 Piano Technicians

A good piano technician can be the recording team's best friend, and if financially possible, it is a huge benefit to have a technician on call during any recording session involving piano. Not only will they maintain perfect tuning throughout the recording, but they can also be called upon to address issues with the action and noises caused by the pedal mechanisms, or to adjust a note that might be sticking out or slightly weak in its production of sound. Many problems such as instrument noise are more effectively addressed by the piano technician than by the engineer. Trying to avoid mechanical instrument noises by repositioning the microphones is usually unsuccessful, and may compromise the overall sound in the process. Having several piano benches available will offer the artist a choice and can also help avoid the use of a creaky stool.

21.6 Chapter Summary

- Listen to the piano at the piano, before positioning the microphones.
- Two spaced omnidirectional microphones may be all that is needed to properly capture a concert grand piano.
- Maintain audience perspective of the instrument, rather than that of the player.
- Having a competent technician on call can greatly improve a recording.

Reference

1. Haigh, C., Dunkerly, J. and Rogers, M. (2021). *Classical Recording: A Practical Guide in the Decca Tradition.* New York: Routledge.

Chapter 22

Recording Classical Crossover Projects

Classical crossover, as a hybrid genre, requires a hybrid approach. Classical artists and labels will expect a certain amount of familiarity of sound, even though the project itself may be a significant creative departure from the artist's usual circumstances. There will be other considerations that come into play based on the blending of various genres, such as a classical soprano singing jazz standards or Christmas songs. In cases like these the singer may prefer hearing their voice placed within a concert hall acoustic setting, whereas the rhythm section will need to be presented using the familiar sound and balance expected of a jazz recording.

22.1 Microphone and Stage or Studio Setups for Crossover Projects

Stage setup and microphone placement will depend heavily on the style or genre of the music to be recorded. For instance, if the piece is an arrangement of a classical work for soloist and electronics, it is quite possible that the soloist will still be presented in a "classical" or traditional setting. This would be implemented through the use of a main system and associated spot microphones, as described in previous chapters of the book. The treatment of electronics might be quite different, unless the composer intended for these to be reproduced in the recording space – although usually electronic soundscapes are presented with very little natural acoustic interaction.

In the case of a string quartet performing arrangements of popular music songs, it may be more appropriate to present the group in a more intimate way because of the style of the musical genre. A more intimate placement of the main system and heavier use of spot microphones in the mix may provide the desired perspective and timbre, or the traditional concert hall may be abandoned completely in favor of the clarity and dryness of a medium-sized recording studio. In this instance, the main sound of the ensemble may very well be derived through the use of spot microphones rather than a main capture system. This is not to say that the resulting recording will be drier than a typical string quartet recording. Instead, more control and detail will be present in the overall sound, and with the addition of artificial reverberation the traditional concert hall setting can be recreated after the fact.

For recordings of this type, it is still strongly recommended that a main pair or an L/C/R system of microphones be set up and recorded, even if it is not intended to be used as a main pickup in the traditional sense. One might be pleasantly surprised at times to find that a main system placed in close proximity to an ensemble might sound more blended and balanced than a combination of close microphones used to record each instrument separately.

Regarding the choice of spot microphones, large-diaphragm cardioid microphones might be an interesting choice, given their wide polar pattern and off-axis high-frequency roll-off. In a situation where a main pickup is not in play, a helpful amount of "glue" or blend will be welcome,

DOI: 10.4324/9781003319429-27

provided here by the leakage of sound from adjacent instruments into each microphone. Ribbons may also be considered, due to their forgiving nature when placed in close proximity to strings and other instruments. The relatively narrower bidirectional pickup pattern of a ribbon may offer better separation between instruments, as needed. The figure-8 polar pattern may also help in providing additional blending of the instruments in space, potentially due to some pleasant room reflections present on the active backside of each microphone capsule.

In a studio setting, the main system may still be comprised of omnidirectional microphones, depending on the size and quality of the room. Larger studios and rooms with pleasant-sounding acoustics will integrate nicely when using "omni" microphones, whereas smaller or less-interesting-sounding rooms may require the use of more directional polar patterns for the main pickup. It is worth doing some experimentation not only in this regard, but also regarding the physical placement of the main system.

EXERCISE 22.1 RECORDING STRING QUARTET IN THE STUDIO FOR A CROSSOVER PROJECT

This approach can be applied when a more intimate sound is required by the genre, whether in a studio setting or on stage.

1. For string quartet, set up four directional microphones less than 1 m or 3 ft. from each instrument in the group. A bow length may be a good indication of a close placement.
2. Add to this a main system up and over the group at a fairly low height. Start at around 2– 2.4 m (6–8 ft.), depending on the size of the studio, and around the same height or slightly higher for a stage setup in a hall.
3. Back in the control room, prepare an appropriate reverb that works well with either microphone setup. As the group plays live (or after a take has been recorded), listen to the close microphones alone, then slowly fade to a mix of 50/50 between close and main microphones. From this point, continue slowly to a mix where only the main microphones are heard. Creating a group fader for each set of microphones will make this comparison much easier to realize.
4. Continue listening back and forth, favoring the close system or the main system, or a blend of the two systems combined, finally deciding on the mix that works best for the style of music being recorded.

When recording jazz ensembles for a classical crossover project, the bass and drums will probably be captured using techniques that are typical for the jazz genre. Piano may be recorded with the microphones placed slightly farther away from the hammers or slightly outside of the instrument, so as to present a more "classical" piano sound – a warmer, more blended sound with a less aggressive attack. In this way, segments of a piece that feature vocal and piano only may suggest to the listener a classical vocal recital soundscape, until the bass and drums join in later in the arrangement.

22.2 Recording Crossover Projects in the Studio

Recording process: in terms of the actual style of recording, it makes the most sense to stick with a typically classical workflow, recording multiple complete takes of a piece or movement and then editing between the takes after the fact. Most classical musicians will not be familiar with popular music production methods such as "punching in" or recording each instrument separately. While the use of a click track is a rarity in crossover recording, checking a metronome before beginning each take is a good practice. This helps to assure that editing between takes will be possible, assuming the ensemble performs each take at about the same tempo. Comparing the duration of each recorded take can also be a good indicator of varying tempo and the possibility of editing between takes.

Isolation booths: most classical artists prefer to perform live on stage or in the studio with the other musicians, without the use of headphones and a "cue" mix. Using an isolation booth may present major issues if the appropriate preparations are not met. If and when isolation is necessary, the soloist will require a high-quality headphone mix that is constantly monitored. This way the producer and engineer can know exactly what the artist is hearing at all times. Knowing that an artist with little experience in studio work will be nervous and uncomfortable from the start, the recording team should be extra prepared. A comfortable booth with good temperature and lighting, a stool or chair ready, a tray for their personal items and water bottle, and even a selection of different styles of headphones, should all be ready to go.

Headphone cue mixes: preparing the headphone mix ahead of time is critical. This may be hard to do for a live studio session, but if there is time to have the ensemble play before the soloist arrives, the headphone mix can be set up properly before the first real take. Artists that are not used to being in the studio will not be comfortable making their own mix on a headphone cue mixer. Best practice is to provide a full mix and to make any necessary adjustments from the control room. Where singers are involved, a good amount of reverb should be ready to go, as this will help place their voice in the familiar, more natural acoustic of a large space, and the sustain provided by the reverb will help them manage their intonation. Hearing the "band" or accompanying instruments well is also a major consideration. Great care should be taken in compressing a classical singer's vocal track. However, a good amount of compression might be useful on the backing track elements in the headphone mix, so that the singer can clearly hear softer accompanying passages and at the same time won't be blasted by louder moments in the track when the headphone volume is set too high.

Be careful when using compression on a classical vocalist's headphone mix. Many singers cannot stand the effect of compression on their voice, especially while they are singing. The common complaint is "why does my voice get softer whenever I sing louder?" Any compression applied should be done in a conservative and transparent way, and careful monitoring of the soloist's headphone send level should be done live on a fader by the recording engineer. The simplest approach is to monitor the mix that is being sent to the vocalist and adjust it on instinct and through direction from the singer. Written notes can be taken in order to repeat certain fader levels over multiple takes for consistency.

Overdubs: setting up a suitable headphone mix for overdubbing can be done ahead of time; if the artist can hear both the track and themselves very well, this should help put them at ease. During the overdubbing process, there will be a need to start and stop at points within the piece other than the very beginning. The operator of the DAW should have a great number of markers prepared, including section names and various significant measure numbers, so that the producer or engineer can have the operator "cue up" to a series of potential starting points. This should be done well before the artist arrives. In the case of the soloist beginning a piece, or in order for them to enter in time with the group after a long pause in the music, certain "count offs" may need to be added. For example, a piece in 4/4 time might require that four clicks be prepared at the incoming tempo and placed before the first entry.

EXERCISE 22.2 PREPARING A HEADPHONE MIX FOR A VOCALIST

This should be prepared well before the artist arrives:

1. In the control room (or backstage) prepare a good starting mix of the backing track. If the plan is a live but isolated vocal, have the accompanying musicians record a test track that can be used to set up the headphone mix. This can be done even before the final mic placements are set.
2. Go into the booth (or on stage) and listen to the mix over the headphones after setting a suitable starting volume.
3. Ask someone in the control room to adjust the balances as needed. Make sure the amount of compression on the accompanying track is working well.
4. Try singing along (or talking) on the microphone to get a sense of the balance between the track and the vocal. Make sure a decent amount of reverb is ready on the vocal channel. In the case of other instruments such as violin, singing or at least talking/scratching/tapping the microphone will provide some indication of the level of the soloist against the track.

Communication: any artist that has been confined to an isolation booth will need clear efficient communication with the control room as well as the other musicians on the recording session. It is the responsibility of the recording team to add "talkback" microphones in each isolated space, and to be in charge of faithfully opening and closing these microphones between each recorded take so that communication is as seamless and clear as possible. A breakdown in communication can risk the demise of an otherwise successful recording project, if and when any frustration sets in.

22.3 Recording Orchestra for Crossover Projects

Crossover projects involving orchestra may not need much change of approach from that of a normal orchestral recording. The biggest differences may simply be the "closeness" of the ensemble capture and the fact that a soloist, whether singer or instrumentalist, may be presented at a much higher level in the mix, as is common in popular music production. A more intimate perspective can be realized by reducing the distance and height of the main microphone system, and with heavier use of spot microphones, as described in Section 22.1. Soloists may be presented using more of a pop music style. Lead elements will be balanced so that they are heard out in front of the orchestra, with a good amount of compression and possibly brighter timbre than a traditional classical mix.

Recording pops orchestra is another challenge altogether. Full orchestra with rhythm section can be very difficult to manage and balance properly. The main culprit will be the drum kit, and although some transparent Plexiglas dividing screens or short solid baffles might help for control, the best scenario is a drummer who understands the complexity of acoustic balance on stage.

Drummers who play too loudly will end up in all the microphones, and the only way to mask the bleed will be to increase the drum tracks in the mix, resulting in a drum-heavy recording. The best drummers in a pops orchestra will play at a more subtle level, knowing that the engineer can always make them louder in the mix, and with clarity and control. Other sections of the orchestra may need more support for clarity and detail because of the rhythm section. String and woodwind spots will be needed in a more important capacity, to bring out lines against drums and brass.

I once worked with a great drummer during a recording of the Bernstein *Mass*. This is a piece with full orchestra, multiple solo voices, three choirs, rock/jazz band, organ, banjo, and so on. As part of preparing for the recording we were placing some baffles around the drum kit to help control the leakage, and he said, "Don't worry, I know how it works – the louder I play, the more you turn me down!"

22.4 Chapter Summary

- Crossover projects tend to combine two or more genres, and with that, multiple textures and perspectives need to be blended together.
- Great care is needed with classical artists coming into a studio setting – headphones and isolation booths may be a new or rare experience for the performer.
- The use of dynamic range controllers on classical singers should be applied with extreme transparency, as most singers do not like hearing any unnatural gain reduction.

Part VI

Postproduction

Chapter 23

Editing

As soon as more than one performance of a piece has been recorded, any amount of editing can be done. These performances can be from multiple "takes" recorded in the context of a studio session, or from several live concerts of the same program over consecutive nights. Cutting back and forth every few measures between two nights is not unheard of in modern recording, as computer editing systems offer seamless splicing with level matching and many options for crossfade lengths and shapes. Copying a few measures from one section of a piece and pasting over the same music in a repeated section is a very common practice in music production, and was so even in the days of analog tape and razor blades. This chapter includes suggested techniques for marking the score with the preferred takes, organizing source audio, arranging the computer editing system for the most efficient workflow and organizing hard drive storage and backups.

A colleague once told me that *"the best editors' work is never heard"* – meaning that if the editing process is well executed, the result will be a seamless combination of all the best moments across multiple takes, yielding a perfect performance. "Seamless" means that there are no editing artifacts such as double attacks on a bad splice, tempo mismatches or sudden changes in level at any splice point.

23.1 Choosing Takes, Marking the Score

During the recording session or live concerts, the producer will be busily documenting both exceptional moments and questionable performances issues. This will mostly consist of marking a minus sign in the score whenever a certain part is poorly executed – due to wrong notes, missing notes and sections or chords where the ensemble is out of tune or rhythmically not together. The goal is of course that wherever a minus sign is written in the score based on one performance, a plus sign should be notated in that same place from another performance, thereby ensuring that an edit plan can be constructed that is free of any of the previously mentioned faults. So, if the brass were badly out of tune in the first take of measure 58, and then in the second take the same measure was perfect, the score would be marked with a "-1 int (intonation)" and a "+2" at that spot.

After a few complete passes at recording a complete movement, the producer may ask the conductor for a few "inserts", or "pickups", to cover any spots that were never performed perfectly. So, if the brass happened to be out of tune in measure 58 on every take, the producer might ask the orchestra to play from measure 56 to 60, or a segment with a suitable starting point just before

DOI: 10.4324/9781003319429-29

the measures in question. The conductor and musicians will be tasked with performing the short section with the same intensity, tempo and character as they performed it with in the main takes, so that the sound and interpretation will match as closely as possible at the proposed edit points.

Editing classical music when the musicians are recorded while performing together in the same room means that all the separately captured microphone signals or "tracks" in the editing system must be cut from one take to another at the same place in time. This is because of the leakage from the various sources into adjacent microphones – for instance a loud brass section will be well heard in the main system and the brass microphones, but also in the woodwinds, strings and harp capture. Therefore, if "take 2" is used to fix a bad brass note, that take must also be an acceptable performance for all other sections of the orchestra at the point where the bad brass note was played. Rather than the previous scenario of measure 58, let's imagine that the brass were out of tune in take 1. In take 2 they were well in tune, but the strings were not together. In this case, another take would be needed to cover that measure, to ensure that every section was performing properly.

It should be quite obvious at this point that the producer needs to follow all sections of the orchestra at once, noting any ensemble issues in the string sections while at the same time listening for intonation in the brass, and so on. In addition, the producer will need to keep in mind the tempo and basic interpretation of previous takes so that the editing between the various takes will be uniform. This can be mentally exhausting for an inexperienced listener – the producer must constantly monitor the performance for intonation, ensemble and overall execution that is musical and appropriate for the repertoire. Clearly, there is no opportunity for idle conversation, and most producers will demand complete silence in the control room when the orchestra is playing so that they may focus on the audio playback. Full concentration is required from the very beginning of the first take, so that all the details of the performance can be observed and duly noted.

23.2 Editing Preparation

It is helpful to add markers in the DAW during the recording session – at least the take numbers, and repeats, if possible. This will make it easier to quickly find each take during the editing process. Also, the name of each audio file should include the take number, so that it is clear which take is in use at any moment in the edited assembly, in a clear visual representation on the screen of the computer workstation being used for editing. An example is shown in Figure 23.1, using the Pyramix workstation from Merging Technologies.

Cross-platform work: platform exchange can be aided by file interchange formats such as OMF or AAF. The idea is that tracks, edits, markers and other information such as mix automation can be transferred between different DAWs that support the format. While AAF-style transfers are sometimes very successful, there will be times when a "bare-bones" transfer is the failsafe option, with no risk of losing any audio segments, for example.

When moving from one platform to another for the editing or mixing stage, it is good to plan ahead. For example, certain software running on a Macintosh computer may default to sorting tracks recorded on a PC in alphabetical order. To maintain the original track order, the files from the session should be recorded with the channel number at the beginning of the track name, to allow the tracks to be sorted the same way on any platform – for example, 01_Left Main, 02_Right Main, 03_Timpani and so on. Including the channel name in each file also helps to keep the audio organized and easily recognizable for another engineer. Be sure to use a double-digit nomenclature when recording more than nine channels, otherwise the track-sorting functionality will not work.

The same applies to changing workstations after editing for the mix. In order to preserve the editing, the tracks will need to be rerecorded as complete files, including all the crossfades and gain

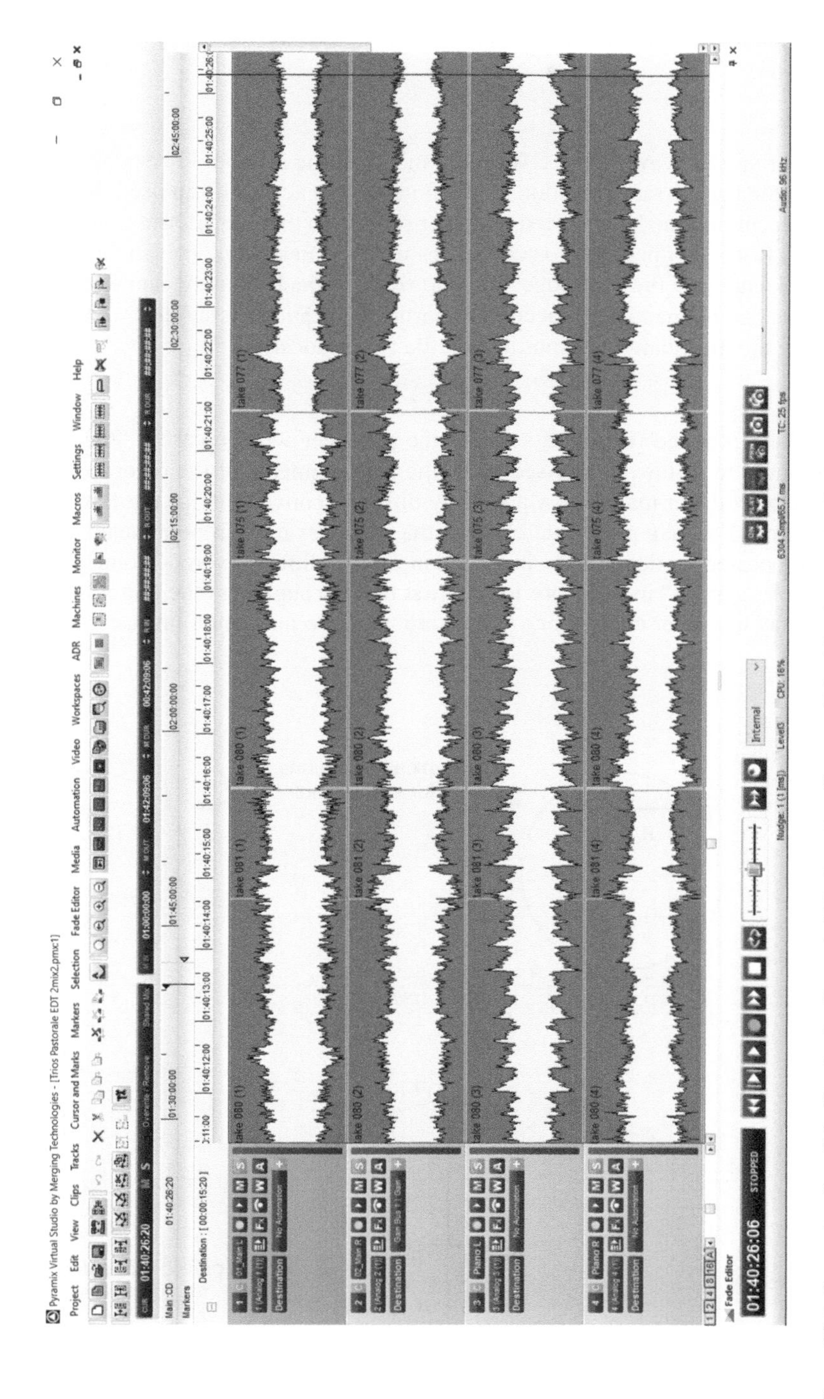

Figure 23.1 Example of an Edit Plan in the Pyramix Workstation Showing Take Numbers in the Filenames

Credit: Image courtesy of Merging Technologies

adjustments between edit segments. This process is typically referred to as rendering or consolidating. Before rendering a multitrack edit, be sure to include a number at the beginning of each track name, so the files will open in the same order for mixing, such as "01_main_left, 02_main_right, 03_bass," and so on.

Rendering as broadcast wave files (.BWF) will maintain the original timestamp, so that the files can be opened at the same location on the timeline of each platform used. In this way, a list of marker locations can be exported, saved or printed so that the information can be used as a reference. If it is not possible to render files with a timestamp, one can simply include the time-code location of the "left edge" (or start time) of the files in the name of the render, so that the files can be manually placed at the original location on another workstation, by simply reading the timestamp and spotting the file to that location.

Based on the takes from the recording session, the producer will prepare a "roadmap" or edit plan for the engineer to follow, comprised of all the best segments of the recorded material (see Figure 23.2). Each producer may follow a slightly different convention for marking the score, so a discussion will need to take place that clarifies the practices used in nomenclature. A separate splice plan for a repeated section may be indicated in several different ways – some producers may use the top of each "system" in the score for the first time through, and the bottom for the repeat, whereas others may use color coding, or a "T" figure with two horizontal lines.

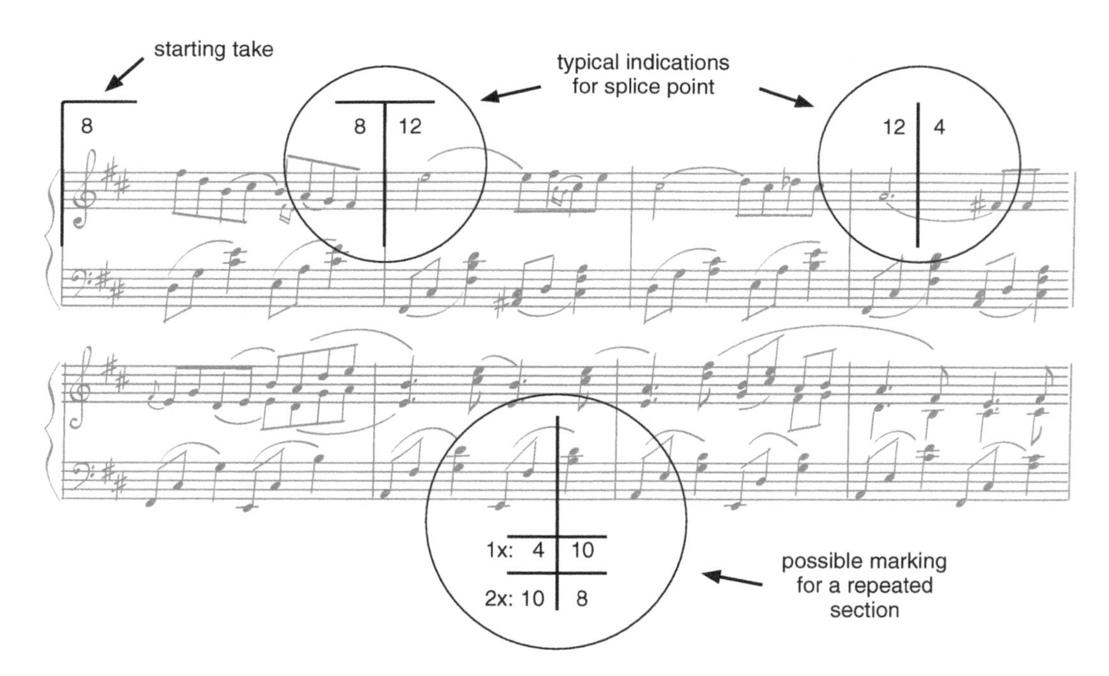

Figure 23.2 Diagram Showing Various Ways to Indicate a Splice Point

Figure 23.3 Diagram Showing Edit Markings with Some Flexibility in the Actual Placement of the Splice

Generally, a vertical line is drawn at the splice point with a number written on each side, noting the "outgoing" take number on the left side and the incoming take on the right. A "T"-shape is more common than a single vertical line, and in some cases an arrowhead may be added to either side of the "T" to indicate that the engineer has a few options as to where in the music the splice may be performed. This can be helpful in "hiding the splice", as the least noticeable placement might be found through some experimentation. In Figure 23.3, for example, the arrowhead on the right side of the T lets the editing engineer know they might splice later, but not any earlier than the vertical line, presumably due to a fault in the incoming take just before the splice point.

Depending on how much latitude the producer will allow the engineer in the editing, the + and – markings in the score from the recording session can be used to show the limits of how far a splice point may be moved around, as the editor may need to find a spot where the two takes are more similar in level and balance. Of course, the simplest interpretation would be to avoid any measures that are marked with a minus sign for that particular take. The amount of discretion afforded to the editor by the producer will depend on the editor's musical experience and the rapport between producer and editor that has developed over previous projects.

23.3 Working Efficiently

Editing can be a tedious and time-consuming process, so it is important to be organized with a suitable methodology. Any time-saving strategies that can be used will add up, even if only a few seconds are saved. Using keyboard shortcuts instead of the mouse will save hours over the course of a project – editors should try to avoid reaching for the mouse as much as possible. Some platforms allow the user to assign their own keystrokes to each shortcut, which allows for great flexibility and the customization of the software. For instance, to listen to the incoming take up to the splice point, one might use the keystroke "I". The letter "O" might be used for the outgoing take, and "E" might be used to listen to the edit over the splice point. These are easy to remember, but there might be a preferred set of keys to use, based on the physical layout of the keyboard. Other platforms have fixed key commands that must be learned. There are various forums on the Internet for all of this, and it is easy to find a lookup table for platforms with fixed key commands. It is worth investing the time in learning these shortcuts, as they will improve the efficiency of the workflow.

23.4 Listening While Editing

Looking back to the focus of Chapter 2, critical listening is the most important skill required in audio editing. This may sound obvious, but it should be stated that editing requires very careful and focused listening, otherwise the edit points will risk being exposed for what they are – joins or cuts between different performances. Proper monitoring in a quiet location is paramount, and the editor should be ready to take regular breaks so that their full attention can be applied while working. Monitoring using high-quality headphones is a good solution, but the edits should be checked on loudspeakers as well. In order to ensure that tempos and volume levels match over a splice, the edit should be auditioned from far enough back before the splice point – two seconds of pre-roll might be a minimum amount while assembling the edit plan. Once complete, the entire piece should be auditioned for any slight differences in level or tempo at each crossfade point.

While adjusting incoming or outgoing levels is common practice in audio editing, small gain changes of 0.5–2 dB may be enough to make a splice between two performances sound seamless. Unless an incoming take is consistently louder than an outgoing segment, any gain change should be "normalized", meaning slowly returning to 0 dB from the adjusted level over a few seconds of music. In other words, if a crossfade on the middle of a phrase seems to bump up slightly in level, the incoming take can be lowered slightly at the splice point, then slowly returned to its normal level over a period of time so that the change is unnoticed. This will keep the edited project close to its original levels throughout.

Unlike most pop recordings, which are often referenced to a metronome or click track, the audio in a classical edit can be slipped around in time to match up splice points, as long as all tracks are grouped together so that they move as a unit. In this way, slight variations in timing or tempo can be compensated for. Also, when a certain event in a take is not quite played at the same time by all the members of an ensemble, the audio can be "clipped". This means that a very small amount of time can be removed between the earliest attacks and those that may come slightly afterward, thereby bring the two attacks almost completely together, to sound as if they were played at the same time. Clipping needs to be done with great care and with the right amount of crossfade time, otherwise it may sound like a hiccup in time, or too much like a robot playing with perfect precision.

23.5 Source/Destination Editing

Many DAWs or editing platforms have a beneficial functionality known as "Source/Destination" or "Source to Destination" editing. Rather than copying and pasting material together to make a composite edit plan, each source audio segment is identified by using "gates" – normally a left and right gate denoting In and Out positions at the beginning and end of a particular segment. The same place in the music is then identified in the destination with In and Out gates, creating what is known as a four-point edit. The edit is then made using a keystroke that essentially activates a macro that copies the gated source material and pastes it into the gated area in the destination. It then applies default or predefined crossfades at the In and Out points of the edit. Any slight difference in length between the source and destination segments is automatically compensated for, as the remaining audio to the right of the inserted segment in the destination is moved slightly

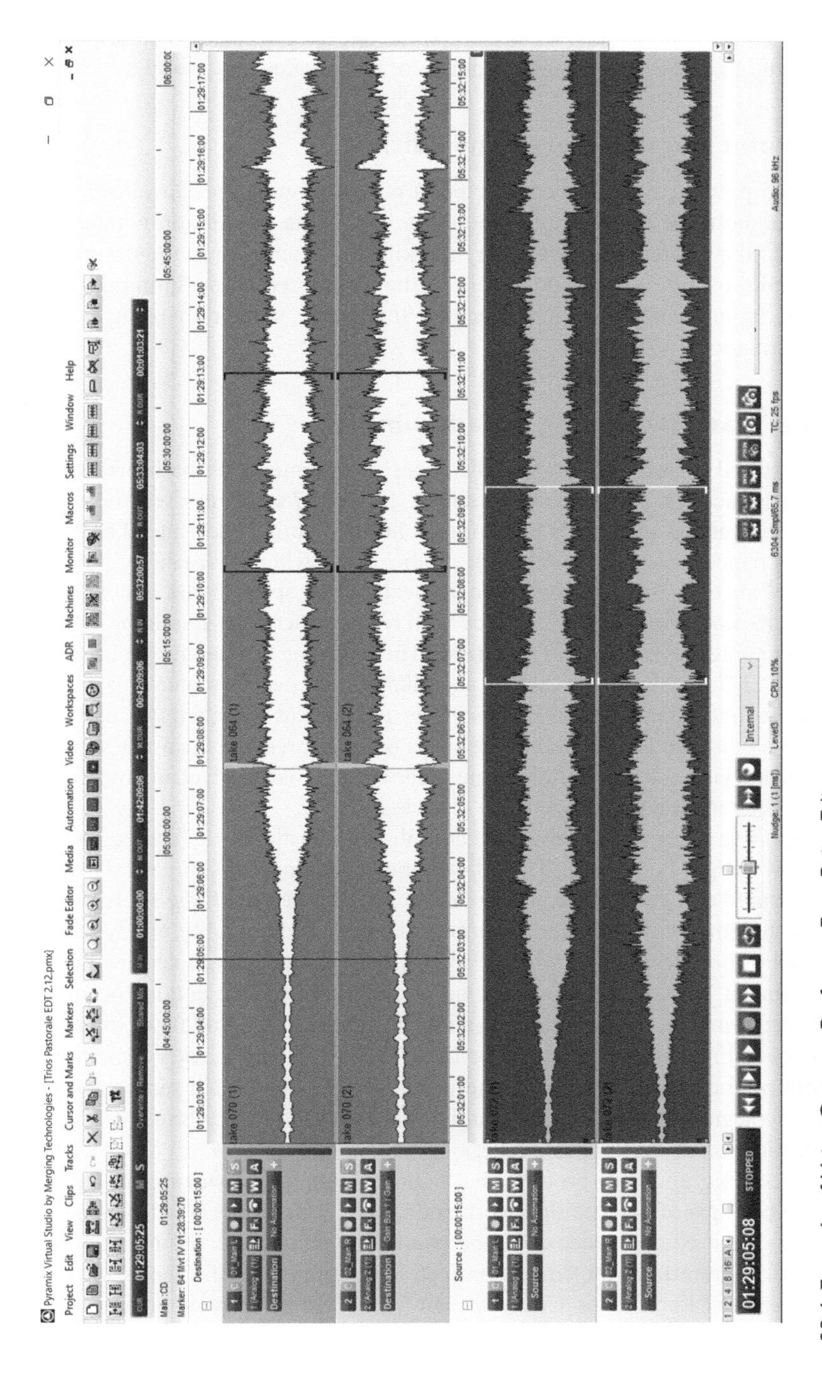

Figure 23.4 Example of Using Gates to Perform a Four-Point Edit

Credit: Image courtesy of Merging Technologies

earlier or later as needed. Multiple segments will be moved together, preserving any editing that has already been completed later in the piece or movement. DAWs that include source/destination editing as a time-saving functionality offer a major advantage over those that only support "copy/paste"-style editing. Figure 23.4 shows an example of using gates to perform a four-point edit.

23.6 Three-Point Versus Four-Point Editing

As opposed to the previously described four-point edit, a three-point edit allows for a segment from the source audio to be inserted into the master edit in the destination without changing the length or timing of the master. This is done by marking an In and Out in the destination and only marking the In point of the source. Conversely, if the length, tempo or timing of the source segment is preferred, In and Out gates can be placed in the source, while only using an In (*or* an Out) marker in the destination.

23.7 Keeping Track (Dropping Breadcrumbs)

As the source material is navigated to find the necessary segments of audio needed to assemble the edit plan, more location markers can be added. If take 57 is being used for a few measures and then will not be needed again for a page or two, an editor might skip ahead and add a location marker (based on page or measure numbers from the score) where that take will be needed next, for instance "Tk57 p. 16". This will save time in locating that spot when it is next needed several minutes later, or possibly the next day. A piece with repeated sections should be carefully marked so that the correct content is used at the right time. The first time through might be marked as "1x" and the repeat as "2x", or simply as "tk57" and "tk57R". Repeated material is often performed differently than the first iteration (as in Bach sonatas and partitas for example), and so it is critical that the producer's edit plan is followed precisely in order to preserve the artist's interpretation of the piece. Hansel and Gretel's idea of "dropping breadcrumbs" is a perfect analogy for the methods engineers use to keep track during the editing process.

When editing a live concert using multiple complete performances as sources, multiple source folders or windows might be created. This allows the "grouping" of sources so that several concerts can be compared and edited between in an efficient manner. With the sources lined up vertically and linked together, locating, zooming and playback functionality are synchronized. This saves a great deal of time, especially when it comes to locating and comparing similar phrases or moments in the music. For example, in a live orchestra recording, the conductor might prefer the Saturday-night performance – except there is a bad horn note in the middle of the last movement. With all the concerts lined up as sources in the DAW, the editor can quickly jump to the Thursday and Friday night performances to see if the horn was played well at that same spot, and then that note (or a suitable segment) can be quickly edited into Saturday's performance, which has been placed in the destination as the "master" take.

Just as the recording engineer protects the audio by running a backup, the editing engineer should also keep backups of their editing work and save editing projects with incremental names, so that any stage of the editing process can be recalled. For instance, the name of the recording project should be different from the editing project, for example "Rach Piano Concerto first night record", as opposed to "Rach Piano Concerto edit V1.1". So as to avoid saving over something worth keeping, each iteration or revision should have a new name (V1.2, or V2.0). It is good to decide on a system with some structure so that the numbers are meaningful. For instance, V1.1 through 1.9 might be versions the editor has prepared on their own, and then V2.0 and later are second-edit revisions based on comments from the producer or the artist. If an alternate ending is created, the project might be named "V1.3 alt ending". All of these edit plans should be saved on at least two separate hard drives or somewhere online (as is done with the actual audio), so that all the data are protected.

In terms of basic "housekeeping" on a hard drive, a methodology should be used that would allow any authorized person to open a hard drive and easily find the most recent version of an edit plan, and to determine which file is in fact the approved edit and what might be alternates to that. Basic strategies here include moving old projects into a folder called "old projects" or "session history", so that only the most recent project is displayed when the folder of a certain piece is opened. This will avoid confusion between multiple editors or between editor and mix engineer, and also serves to remind an editor where they left off if they have to come back to a project after a significant amount of time has passed, as many other projects may have come and gone during that time.

23.8 Chapter Summary

- Be sure to understand the producer's edit plan notation before starting the edit.
- Use shortcuts and key commands as much as possible to maximize efficiency.
- Organize data files and folders in a clear way so that any user might navigate the project.
- Back up the work at the appropriate stages and assign each iteration of the edit plan a new name.
- Check the splice points on both loudspeakers and headphones to ensure the editing is seamless.

Chapter 24

Mixing

This rather in-depth chapter places emphasis on mixing approaches and techniques, as well as tips for saving time during the mixing stage of a project. In live mixing, many techniques are the same as those for mixing after a multitrack edit has been completed – and certain concepts are identical. Use of artificial reverberation is an important part of this chapter, and the use of automated mixing, equalization and dynamic compression in classical music is also discussed. Additionally, mixing to picture is covered, and the practice of matching the audio perspective to the visual image is explained.

"Listen, listen – it's simple. If it's too soft, turn it up. If it's too loud ... turn it down!" – George Massenburg, on the subject of mixing.

24.1 Mix Preparation

As presented in Section 2.4, it is important to listening to the control room before any work begins – whether it is recording or mixing. The engineer needs to be well acquainted with the listening environment before making any decisions affecting balance and timbre. Here, the engineer and producer should audition a few well-known excerpts of recordings that are familiar to both of them, to be sure of what is happening acoustically in the mix room. The priorities during listening are overall frequency response including the presence of room modes, the liveness of the space and ensuring that a mono playback is precisely localized in the center of a stereo pair of loudspeakers. This is most important for those mix rooms that are new to the recording team, but also for rooms that may have been used in the past, but with different loudspeakers.

Even familiar settings should be checked quickly in the same way to refresh the aural memory and to be sure the monitoring chain is functioning properly; that is, with no distortion or defective loudspeaker elements, and an equal level in the left and right channels. A more thorough test might include checking each loudspeaker driver separately by covering and uncovering mid- and high-frequency drivers respectively. The low-frequency driver is virtually impossible to cover up, but it should be quite obvious if it is not producing any sound when auditioned at close range (and at a moderate volume). The team should also listen from different positions around the room, assuming that the producer might be listening behind the engineer, for the most part.

DOI: 10.4324/9781003319429-30

Carefully check the monitoring system before listening – when using two-way or three-way "passive" loudspeakers with an external amplifier, it is often the case that the jumpers on the connecting terminals behind the speaker cabinet come loose, so that no signal or only intermittent signal reaches the mid/high drivers. For "active" or powered monitors with an onboard amplifier, check that the filter switches and volume controls are all set in the same positions. An easy way to ensure equal output volume from both the left and right monitors is to set the control to maximum on the loudspeaker and adjust the volume accordingly from the mixing console or the output of the computer interface.

As part of the mix preparation, notes from the recording session should be consulted as to preferences in microphone choices and balance between main and outrigger pairs, so that these details can be compared as they are revisited in the mix room. Also, certain important details from the session will need to be addressed during mixing. For example, a note such as "*perc. 5 level significantly lower than perc. 1 through 6*" means that if all the percussion channels are to be evenly introduced into the mix, "perc 5" will need to be set at a higher level than the others. This may seem like an obvious issue, but if there are 48 recorded channels or more, this type of imbalance might go unnoticed when scanning through the channel meters. The levels should be checked by ear as well.

A new set of notes will normally be produced during the mix sessions, documenting reverb settings, loudspeakers used and any other details that may come in handy later on. Saving multiple versions of a mix under separate names (e.g., V1, V2, end of day) can help the engineer remain organized. There will be times when it is necessary to "backtrack", when it is discovered that something needs to be checked from an earlier point in the mixing process, or the producer or artist wants to return to a balance or setting from earlier in the project.

24.2 The Secret of Mixing

Once the consensus has been reached that the mixing environment is fairly workable and "understood", the process of balancing can begin. After presenting the "secret of recording" in Section 6.3, here the "secret of mixing" is discussed, and it might sound rather unorthodox. Balancing or mixing could be more accurately described as *un-mixing*, whereby decisions are made throughout the mixing process to highlight or favor a certain instrument or section over another. In fact, a "perfectly balanced mix" might sound quite boring, as if some elaborate computer software were employed to analyze the volume or loudness of each track and adjust the levels accordingly. Of course, every element in the score needs to be heard, and for the most part, heard clearly, but a certain amount of very subtle "over-mixing" will help a recording sound more interesting. Certain textures and balances might be designed to change in conjunction with the musical program throughout a movement or a complete piece.

In all types of musical program there exists a traditional unbalancing of elements – one of the most common examples is how unnaturally loudly the vocal is presented in pop music. In concerto recording, we are faced with the struggle of managing the dynamic levels of 60 musicians or more, so that a single element may be heard clearly through the complex tapestry of accompanying voices. Invariably, the soloist will need help from the mixing engineer to ensure they are properly "balanced", or actually *unbalanced*, as they are artificially raised in level.

One skill that takes years to master and is especially helpful in mixing is to be able to evaluate small details and individual instrument balances while assessing the "big picture" at all times. This simultaneous monitoring of "micro" and "macro" elements together can be exhausting, but eventually it becomes second nature for seasoned producers and engineers. Casually listening to recorded music and critically assessing musical program for balance, perspective and timbre are two very different activities, and listening to the same piece of music repeatedly requires a great amount of focus and investment. The producer is normally responsible for the final outcome of the recording, and therefore it should be expected that they will attend the mixing sessions and be listening as carefully as – and at times more carefully than – the engineer, asking for many balance changes to be implemented along the way, and closely following the intentions of the composer and the performing artist, as well as their own instincts and preferences.

Left to right, and front to back – maintaining a stable sonic picture in a stereo mix requires that there is equal energy (for the most part) across the lateral soundstage, and that the depth remains constant even when spot microphones are being raised and lowered in level during the mixing. For example, long passages with heavy emphasis on the violins might warrant raising the right outrigger to balance the additional energy on the left. Keeping the back half of the orchestra present but in the right perspective is also critical.

While a considerable amount of concentration is spent on small details in the audio, the engineer should be careful to monitor and maintain the listening volume in the control room at a reasonable level, so as to avoid the short-term effects of fatigue as well as hearing loss over the long term. When the ears become "tired" (in combination with the brain), the best course of action is to take a break from listening, rather than turn up the volume. Ten minutes away from the mix will refresh the brain and give the ears a break, so that the volume doesn't need to be raised. Choosing a reasonable volume at the beginning of the day and sticking with it will provide a standard reference while establishing balances. It is good practice to make a note of the actual level as a number or a pencil mark on the volume control, so that it can be reset if changed at any point. Following this protocol will allow for more productive results over the long hours that are normally associated with postproduction. Young audio engineers hoping to observe a long and successful career will want to protect their hearing at all times.

24.3 Listening to the Main System

No matter what the original impression of the main system happened to be during the recording session, it is a good idea to begin a mix with a reassessment of the main microphones. There are several reasons for doing this, the main reason being the importance of assessing all of the elements separately before combining the channels, so as to create the best possible balance and representation of the recording. Secondly, it is assumed that the mixing control room will offer a more accurate listening experience than the room used backstage at a concert hall or an unfamiliar control room at a studio.

Even when the control room at the recording location has been evaluated, it cannot be assumed that the recorded channels were all optimized in terms of the microphone placement, or that the monitor mix from the session represents the best possible result. Time constraints and the stress of potential technical issues combined with the pressure to avoid leaving a session without having covered the musical content properly first will affect the recording team's ability to critically evaluate the audio at that time. When rushed for time, one might accept the metric "good enough" over the more desirable standards "good" or even "perfect". During the postproduction stage, an appropriate amount of time can be allocated to careful listening and assessing balances, and small adjustments can be made to improve the overall quality of the recording without overly obsessing about how much time is being used.

24.4 Introducing "Spot" Microphones into the Mix

Once the outriggers, woodwinds and other channels have been introduced to some degree into the presentation of the main pickup, the actual "balancing" or mixing is fully underway. Unless a second main system is in use, the next logical set of channels to assess is the outrigger pair. If close and far pairs of mains are being combined, this balance should be established before proceeding.

Using a single group fader that controls both left and right outriggers, or the actual outrigger channel faders, slowly open them into the mix to the same level as the main pickup and listen to the change in sound while adjusting the level. The section blend of the strings should improve with the outriggers open; the overall image will be wider, and the strings should be more present in general. Once a satisfactory level is established, it should be noted that the outriggers might still be raised and lowered from this departure point for various reasons throughout the mix. As discussed in Chapter 7, the string section spots might not be needed as part of the basic makeup of the balance – in which case they may be reserved for the occasional solo, or a section in which the brass are dominating and a small amount of "bite" is missing from the strings, for instance in a large tutti section.

"*Less, in fact ... is less*" – is a realist's interpretation of the "less is more" adage. A very good recording can easily be ruined in the mixing stage through excessive use of the spot microphones as compared to the main system. According to the "secret of recording" in Section 6.3, it would be naïve to simply open up all the recorded channels into the mix just because they exist. If the main pickup has been properly positioned, the supporting microphones should not be needed to a great extent, and it might be the case that very few are needed at all in the final balance.

The woodwind section might be considered next, as quite often a nominal amount of signal from the woodwind channels may be needed in the mix as a constant to keep the overall depth of the ensemble in check. It is important to remember, however, that if the woodwinds are loud enough in the main system and the depth of the ensemble is not overly exaggerated, the spots may not be needed all the time. Featured musical lines and solos may still require some help here and there.

Brass instruments occasionally present themselves as too bright in the main system and may need a small amount of spot signal in the mix to "round out" the sound, help to focus their stage position or to highlight a solo. French horns, with their rear-facing bells and warm character, generally need a greater amount of focus as compared to the trumpets and trombones, which are directly facing the main system.

Percussion, piano, harp and celeste channels are only needed when they are playing and should be removed or lowered in the mix when not needed. It is common with timpani that only a small "point" is needed to complement the sound of the drums in the main pickup, and a high pass filter (HPF) might be applied so that the low-frequency content in that channel is removed, and only mid- and high-frequency energy is used to add clarity to the timpani sound in the mix.

24.5 Fader Automation

Once a satisfactory "static" balance has been established, with all the faders set at stationary positions, dynamic mixing, or automated mixing, may proceed [1]. This involves listening through the piece and correcting balances phrase by phrase, as needed. In this static mix, some channels may be completely closed or off, only to be opened for certain sections or phrases. The resulting automation might be quite sparse, or there may be various faders moving almost all the time, depending on the repertoire. Concerto mixing will normally entail the soloist's level being adjusted almost constantly, even if it is within a very small range of motion on the fader.

Using the mute or "channel on" switches in classical mixing is much less practical than simply opening and closing faders. This is because there is a great deal of leakage in every channel that is recorded as part of each channel's "main" signal, as the musicians all sit on stage and play live together, for the most part. A good example would be an English horn solo following a loud brass chorale. If the English horn channel is "unmuted" in the middle of a brass chord, there will be a noticeable bump in the brass level, along with fairly obvious timbral and image shifts.

> The instantaneous "unmuting" of a channel in a mix will cause an abrupt change in the amount of ambient sound, because of the inherent leakage of adjacent sources in each microphone on stage. Carefully opening a fader over a certain period of time will result in a more subtle and "musical" introduction of residual background sound coming from neighboring audio sources.

To return to the concept of simultaneously tracking micro and macro, it may be decided that halfway through the process of automating the faders that the outriggers, for instance, are not at the optimum level. At this point they may be trimmed up or down slightly for the entire movement or piece, in a separate adjustment from all the "local" fader cues or moves.

24.6 Dynamic Processing in Classical Music Mixing

While compressors and limiters are rarely used in classical music mixing, their use will be required from time to time, and as such their function should be well understood. The exceptions here are extremely dynamic content, or classical "crossover" projects, wherein classical artists step out of their comfort zone into genres such as jazz, pop or Latin music. For more background on the fundamentals of dynamic range controllers, see [2] and [3]. The main conflict with dynamic range controllers and classical music program is that the artifacts are much more obvious than in pop music recording. Every channel has a minimum amount of background information, which is a combination of reverberation and leakage from other instruments. As soon as any gain reduction is applied to a channel, the leakage will also react to the processing and will generate some disturbing side effects. Assigning a compressor to the mixing buss can be very dangerous and must be done with great skill and attention, and only when absolutely necessary.

There is nothing more anticlimactic in classical music than to experience a crescendo that builds in intensity over several long phrases, only to be struck down by the entrance of the timpani, as the compressor "kicks in" and causes the strings to virtually disappear in volume.

Pieces with excessive dynamic contrast are managed more effectively with manual level control. For instance, a movement with an incredibly loud beginning and ending may benefit from having a very soft middle section raised in level, which can be done subtly and over a period time so that it doesn't sound as if someone is manipulating the volume during playback. The result will be greatly preferred over the use of a compressor on the loud passages.

24.7 Equalization

Equalization is the practice of modifying or refining the spectral content of the audio signal [3, 4]. To a certain extent, very little equalization should be needed in mixing classical recordings. In the days of analog mixing, processing was kept to a minimum to avoid the added noise and degradation of sound caused by having too many processors in the signal path. Now that digital signal processing (DSP) has advanced to the point of virtual transparency in operation, it is less of a concern, although high-quality microphones positioned in the right places should negate the need for much timbral correction in the mixing stage. Equalization should be thought of as a last resort, where simple balancing and possibly some added reverberation should yield an excellent result. Filters are helpful tools, however, for cleaning up spot signals, as previously mentioned in the case of timpani. In any case, it is worth listening to a channel before a filter is applied, to see if it is really needed.

24.8 Artificial Reverberation

Artificial reverberation is a remarkable tool for enhancing recorded sound. Adding reflections to a source signal will increase the sense of power and richness of the sound. This is because an increase in reflected sound from a signal gives the impression that it is quite loud, as it is exciting the surrounding environment in a stronger manner. Adding reverberation is a great way to increase perceived loudness without adding gain. As stated in Section 6.4, incorporating room microphones into a mix will provide additional reflections and diffuse energy, but for better or worse, the quality will be much the same as in the main microphone system. This is especially true when omnidirectional microphones are used. For a significantly increased decay time, or longer "reverb tail", an artificial reverberator will provide a more suitable result in a shorter amount of time.

Just as spot microphones should not be brought into a mix without careful consideration as to their level and function, reverberation should be used in a very specific manner, and the result should be critically evaluated. Using the wrong preset, for instance, can be catastrophic. While subtle amounts are easy to manage, it is also possible to add a significant amount of reverb that "complements", as opposed to exactly matching, the sound of the recording venue.

The trick to working with reverberators is to adjust the parameters so that the added reflections and dense decay are seamlessly blended with the original sound. As long as the listener is unaware that anything synthetic is influencing the recorded sound, the engineer can dramatically improve an otherwise uninteresting or exceptionally "dry" recording. If the reverb sounds artificial, there is either too much of it, or the settings have not been optimized to the fullest extent.

Reverberation in mixing should be managed by way of an aux "send" or buss, rather than by simply adding reverb to the finished mix [1]. This will allow for more precise allocation of spatial cues, and the end result will be more natural. If a recording is lacking in spatial information, or a longer decay time is desired, adding reverberation to the main system will greatly help. Where the spot microphones are concerned, the amount of their signal sent to the reverb aux buss will depend on the circumstances. A particular source that lacks focus or needs "drying up" in an otherwise-reverberant main pickup will benefit from less signal being routed to the reverb channels. Instruments with a lower range, such as string bass, timpani and tuba, should be treated conservatively in order to avoid an unpleasant buildup of low-frequency reverb in the mix. In cases where an instrument needs strong reinforcement, a liberal amount of reverb will be needed from that specific channel in order to smoothly incorporate the signal into the overall balance. Very dry spot microphones, when blended with an appropriately reverberant main mix, risk being overly exposed in the balance, causing a breakdown in the natural perspective.

EXERCISE 24.1 EVALUATING REVERB IN A MIX

There are two ways to critically evaluate the parameters of the added reverberation:

1. Monitor the reverb buss alone while the mix is running – mute the main mix buss and route the reverb return channels directly to the output, as opposed to the mix buss.
2. A less sophisticated method, although faster to realize, is to simply run the mix and stop the playback in the middle of program while concentrating on the character of the reverb decay in the absence of musical program. Adjust the reverb parameters a needed and repeat the process, using the same musical segment or other sections of a piece.

Reverberation parameters: most of the available controls on various reverberators, whether using a software plug-in tool or a standalone device, will be quite similar across manufacturers. The actual settings of each parameter may appear to be quite different from one brand of reverb to another, although the results will be similar. For instance, a reverb "time" of 1.8 seconds may be an average decay time, whereas the low-frequency component of the reverb may actually take much longer to die out. Once parameters such as "shape" or "spread" are adjusted, the overall decay time may also be affected. Pre-delay, which dictates at what point the reverb begins to react to incoming signal, may be the most common control. A pre-delay setting of 15 to 24 ms will give the impression of strengthened early reflections around the stage area. Spot microphones that need to be recessed in the mix when introduced at rather substantial levels may benefit from a pre-delay of 0 ms so that the reverb is activated immediately, giving the impression that the source image is farther back on stage. A "zero" pre-delay on the main microphones can help to mask the character of naturally occurring early reflections when they are deemed to be unpleasant, as in a mediocre hall, or a studio with a rather low ceiling.

Other parameters such as time and size will be variable between devices – this is where listening to the reverb alone will be the most informative practice. In general, different lengths of decay and room sizes should be chosen for each application. The following settings for overall reverb time are offered as a rough guide:

- Chamber music: ~1.8 seconds
- Opera: ~2.0 seconds
- Orchestral repertoire: ~2.2 seconds

These numbers will vary greatly based on the length of the reverb at low frequency as compared to mid frequencies in each program. The reverb size should also be modified (if available as a parameter), following the above formula, whereby a smaller-size reverb is more appropriate for chamber music than an orchestral recording. Timbral control or equalization of the reverb is usually controlled by the parameters "high-frequency cut-off" or "roll-off", or there may be an actual equalizer built into the software. Convolution reverb simulates a real hall or space based on an impulse response (IR) measurement. This may be a desirable option because of the very realistic and natural sound with a highly complex set of early reflections, but the user may find the controllable parameters do not offer enough adjustability. Algorithmic programs tend to offer more complete variability, and therefore virtually any result is possible, which in the end may sound more appropriate when adding reverb to program that already has a specific reverberation characteristic.

24.9 Concerto Mixing: Mixing Orchestra With a Soloist

A thorough assessment of the main system is still a good place to start with concerto mixing, to get an idea of how much soloist is in the main microphones and the character of that capture. The solo spots can be used not only to make the soloist louder or more focused, but also to compensate for any undesired coloration in the mains. For example, a timbral correction might be performed if a solo violin is too bright in the main pickup, by filtering (in this case darkening) the spot microphones in order to correct the issue.

Concerto mixing can be divided into three steps. Firstly, it is advantageous to start mixing without listening to the soloist. Of course, there will be a nominal amount of the soloist's sound present in the main microphones, but this should be ignored at first while the orchestra balance is perfected. It will help to mute the solo spots while this is taking place.

The second step will entail listening through the piece and first deciding on a static level that presents the soloist in the right perspective and at a decent balance to the rest of the ensemble, and from that point the soloist can be raised and lowered as needed to keep them in front of the orchestra but without dwarfing the ensemble. There will be times in most pieces where the soloist gets buried in the accompaniment, and nothing can be done – when auditioning the soloist channels alone and even the orchestra leakage is louder than the soloist's signal, there is simply no point in raising the level. Over the years, composers have written music with this effect in mind, as when the orchestra overwhelms the soloist it generates tension and excitement in the sound. Most conductors and "some" soloists are also aware of this phenomenon.

The third phase of concerto mixing begins once the soloist is properly balanced throughout the piece, using automation tools to record and playback the fader moves. A thorough listening should be performed, during which the balance should be checked for any places in the piece where the orchestra (or sections thereof) might be raised to help cushion an otherwise unsupported solo instrument. This might include a sustained string moment that was played slightly too softly, or a moment between the soloist and a trumpet at the rear of the stage that might need to be raised to improve the balance now that the soloist has been brought forward in the mix.

24.10 Live Mixing

Live mixing is an exciting and challenging task and is a necessary skill to have as a recording engineer. Careful remixes in postproduction are not always possible, as broadcasts happen in real time and archival recording budgets do not normally include a second mix. As mentioned in Chapter 15, a good overall balance can be achieved during the dress rehearsal or soundcheck. From that point it will require "hands on faders", as constant adjusting will be required – especially in repertoire involving soloists.

In some cases, the live mix from a recording session may be used instead of a remix, in order to save time in postproduction. If a producer has given balance cues during the recording session, they may be able to simply provide an edit plan after the fact, and the stereo mix from the session may be edited together and used as the final balance. This technique requires the engineer to repeatedly perform the same mix over multiple takes of a movement or piece, so that the levels and balances will match during the editing process. Taking notes in the score can help, and for a concerto, the level of the soloist should be noted whenever dramatic changes are made. An engineer that can follow the score while mixing live will be a great asset to the producer.

The multitrack recording can still be used for small balance fixes, and in fact simple remixing of small segments can be done quickly as long as the mix from the session is printed with the individual tracks, in a synchronous manner. A request for "more clarinet", for example, may be met by simply combining the mix tracks with a small amount of the clarinet or woodwind channels for the required number of measures. This way the overall sound will match easily, and the new mix segment can be edited into the live mix tracks at that point.

The best way to develop live mixing skills is through practice. Memorizing where each instrument is located on the mixing console or fader surface will greatly improve reaction times to balance changes. Becoming familiar with the repertoire will help up to a point, although the key to a successful live mix is to be listening at all times and correcting the balance with smooth and efficient moves of the faders.

24.11 Mixing to Picture

Mixing classical music for a video production of a live performance can be both a frustrating and an interesting process. There is the added challenge of matching the perspective of the picture, as well as keeping the audio in sync with the video over the various edits performed throughout the production. Generally, the video edit is executed first, after which the audio can be conformed to match the picture edit. An edit decision list (EDL) is usually provided by the video team, which can be followed for the audio edit. As long as the audio recorder and the video cameras were receiving the same time code during the recording, generating a synchronous edit plan of the audio should be fairly easy. For more information on synchronizing audio and video, see [1] and the references listed at the end of Chapter 15.

For the mix, it is important that an overall perspective is presented that fits the look of the video. The picture cut will be comprised of close and long shots, and the audio should maintain some average sense of distance from the ensemble that makes sense, whether the video is showing a long shot or a close-up.

In general, an audio mix for video will be created with a closer perspective than an audio-only recording. Close-up camera shots can immediately bring any instrument to center screen, so it is important not to present too much width in the image except for general string sound. For example, a harp that is placed on the extreme left of the stage might be positioned only halfway between left and center in the mix, so that any harp close-ups presented in the middle of the screen are not disparate with the sound of the harp that is "off-screen".

Balance will need to be adjusted based on choices made in the picture cut, rather than solely following the musical score. Subtle changes to the mix can help to ensure that whatever is seen is also heard, so as to not confuse the viewer. There exists an old idea that whatever instrument is pictured on screen must be raised in level to be heard over all other instruments. Strictly following this guideline will result in a very choppy mix that may not make any sense musically. A better strategy is to make sure any instrument featured in the video is heard clearly, but not necessarily raised to be the loudest element in the mix. For instance, a harp glissando in a piano concerto – if the harp is featured in a close-up in the video, it should be heard clearly, but not raised to the point where it masks other more important elements, such as the piano solo.

One successful practice is to listen through the finished mix with and without the picture running. Once the mix seems to be coming together, it is worth playing through and watching the video, looking for any instances where the audio may need to be adjusted to match the image. A second "run-through" might be performed without the picture running to make sure nothing is overly emphasized, although an experienced mixer can address this concern by looking away from the screen at times, or during questionable passages where it was felt that certain channels might have been raised excessively.

24.12 Mixing Classical Crossover Projects

Classical crossover recordings will require a slightly different approach during the mixing stage. While many of the preceding sections of this chapter also apply to the crossover genre, certain methodologies will need to be borrowed from jazz and pop music production. In most instances, compression will not be used in classical production. Normally the full dynamic range of the music, along with the musicians' naturally performed phrasing, is expected to be unaltered in the final mix. Projects taking influence from pop and jazz genres will need some dynamic range control so that the average loudness is in the same range as other pop or jazz tracks that might be heard in the same sequence or playlist. To that end, a crossover recording of vocal and string quartet will need to be mixed so that it can compete with a jazz track featuring vocal and rhythm section. Vocals will be compressed and made brighter and may need additional treatment such as "de-essing" – the process of reducing sibilance, which is rarely needed in classical music production. Speaking of compression, great care will be needed to ensure that a classical singer still has a natural presentation of phrasing, something that heavy compression may undo in a performance. Low ratios and conservative thresholds will be needed to keep the vocal sounding "natural". Additional gain management will need to be performed by the mixing engineer in a transparent and musical way to achieve a reduced dynamic range without the usual artifacts of heavy compression.

24.13 Headphones and Mixing

Mixing while listening on headphones is convenient and cost-effective, as it offers a consistent and portable listening environment at a fraction of the cost of decent loudspeakers and a suitable listening room. As a project moves from the recording venue to the editing suite to the mix room, having the same headphones available will provide some continuity in monitoring at each phase. However, headphones present center information quite unnaturally in the middle of one's head, and as such evaluating the level of a centrally located source is quite difficult. Also, headphones offer increased clarity in low-level detail, which causes a tendency to generate a mix that is somewhat dry when heard on loudspeakers, as the level of reverberation is frequently misjudged. The best approach is to set balances on loudspeakers and then check the mix on headphones, as many consumers listen with headphones more often than with loudspeakers. Switching back and forth should be a common methodology and can be practiced as described in Exercise 2.3.

24.14 Chapter Summary

- Listen to the mix room before beginning to balance the recording.
- Listen to the main microphones again before starting any mixing.
- The secret to well-executed live mixing is to react quickly, yet smoothly, to any balance issue that needs correcting.
- Keep accurate notes, and use "Save As ..." with incremental versions so as to keep track of various milestones throughout the mixing process.

References

1. Rumsey, F., and McCormick, T. (2014). *Sound and Recording: Applications and Theory*. Burlington, MA: Focal Press.
2. Case, Alex U. (2011). *Mix Smart*. Waltham, MA: Focal Press.
3. Corey, J. (2013). *Audio Production and Critical Listening: Technical Ear Training*. Burlington, MA: Focal Press.
4. Woram, J. (1982). *The Recording Studio Handbook*. Plainview, NY: ELAR Publishing.

Chapter 25

Final Assembly and Mastering

General guidelines are presented here for final assembly and best practice for matching levels between pieces; how much pause time should go between movements and pieces on the same album; and use of room tone, along with discussions of loudness levels in general and the use of compression in classical music when necessary.

Mastering an album of classical music program is rather a different process than for other genres, as the final mix normally addresses all issues, including timbral settings, dynamic range and image width. Compilation albums may require more processing to match tracks coming from various sources and recording venues as well as the different sizes of ensembles. Some classical compilations will be sourced from recordings made over many decades, and the quality of the recordings may vary dramatically. There are several well-written publications that focus on mastering for popular music. Classical music production, however, requires a different approach in the final stages, based on a rather hands-off philosophy.

25.1 Mastering Preparation

Once again, those most familiar recordings will be dusted off and auditioned before any mastering begins. This will be the last opportunity to hear the mixes before the recording is released. Performing the final assembly in a different listening environment than the mixing room can offer certain advantages, but it is not necessary. The monitoring environment should include the use of full-range loudspeakers, so that a proper assessment of low-frequency content may be fulfilled. Recordings made on location may have significant "rumble" or other low-frequency peripheral noise, including passing trucks, heating and cooling systems or general traffic sound. These issues may go completely unnoticed on smaller loudspeakers.

25.2 The Basics of Classical Mastering

The most fundamental step in mastering or assembling a classical album is to make sure the levels are adjusted from piece to piece so that the listener can comfortably play through the entire program without having to change their volume setting. In acoustic recording, it is typical that the recording levels are set so that the loudest musical passages reach the maximum 0 dB point, or slightly under, so that the best possible signal-to-noise ratio is achieved. The microphone gain settings will be much lower for recording full orchestra than for a solo instrument, as the dynamic contrast between loud and soft will be very different. The result will be that the average level for the solo instrument will be much higher than for the orchestra. It stands to reason then, in combining

DOI: 10.4324/9781003319429-31

repertoire of solo instruments with orchestral pieces, that the solo tracks will need to be lowered in level. It wouldn't make much sense to hear the end of a massive symphonic work followed by a solo violin that sounds as big and loud as the entire orchestra in the preceding piece. The listener would need to jump for their volume control to correct the imbalance. This may seem contrary to a "pop" approach, where gain is normally added to each track to achieve a maximum level.

There are two methods of comparing levels from track to track in an assembly. One is listening from the end of a piece into the beginning of the following track and evaluating the transition. The other method is to play the middle of a track for ten seconds or so, then quickly jump to the middle of another track to get an idea of how the overall levels compare. These two techniques should be used in tandem for the best results. Once the levels have been set using these two methods, it is recommended that the entire album be auditioned in one pass rather than interrupting the playback to correct any level mismatches and awkward pause times. It is best to simply take notes along the way, and once an overall impression of the entire program has been made, the questionable transitions may be addressed.

EXERCISE 25.1 ADJUSTING LEVELS AND PAUSE TIMES BETWEEN TRACKS

1. Play the last 30–60 seconds of the first track and without adjusting the listening volume, continue into the next track, noting any shift in overall level between the two tracks.
2. This is also a good time to judge the pause between the two tracks. A general guideline might be around 4–6 seconds between movements of a symphony, and 8–10 seconds between symphonies.
3. The pauses should be adjusted by "feel", depending on the musical content and the manner in which the "outgoing" track ends. For instance, if the first track ends with a strong climax and the next movement or piece begins very softly, it might be best to add a second or two to the pause to allow the listener to "clear their ears" (aural memory). Conversely, if the first piece takes its time to finish, on a sustained chord at a moderate volume, a shorter pause might seem more appropriate before the next entrance.
4. Listening to only the last few seconds of a track will usually make a pause feel too short and may influence the engineer to add too much time.
5. Once the entire album has been evaluated in this way, it is helpful to play the middle of each track for a few seconds, skipping from track to track to see if anything jumps out or seems recessed in level. Small level differences between tracks will be perceived as shifts in overall perspective, to the point where gain changes of as little as 0.5 dB will be made to match the tracks.
6. Some tracks may need slightly different gain changes on the left and right channels, as certain mixes may appear to be left- or right-heavy, especially when paired up with other pieces in an album assembly.

25.3 Filtering or Equalization in Classical Mastering

Using equalization to match tracks should only be considered after the level differences have been taken into account. It is rare to find a track that needs a different timbral setting from the others, especially for album projects that were recorded in one venue with the same microphone setup.

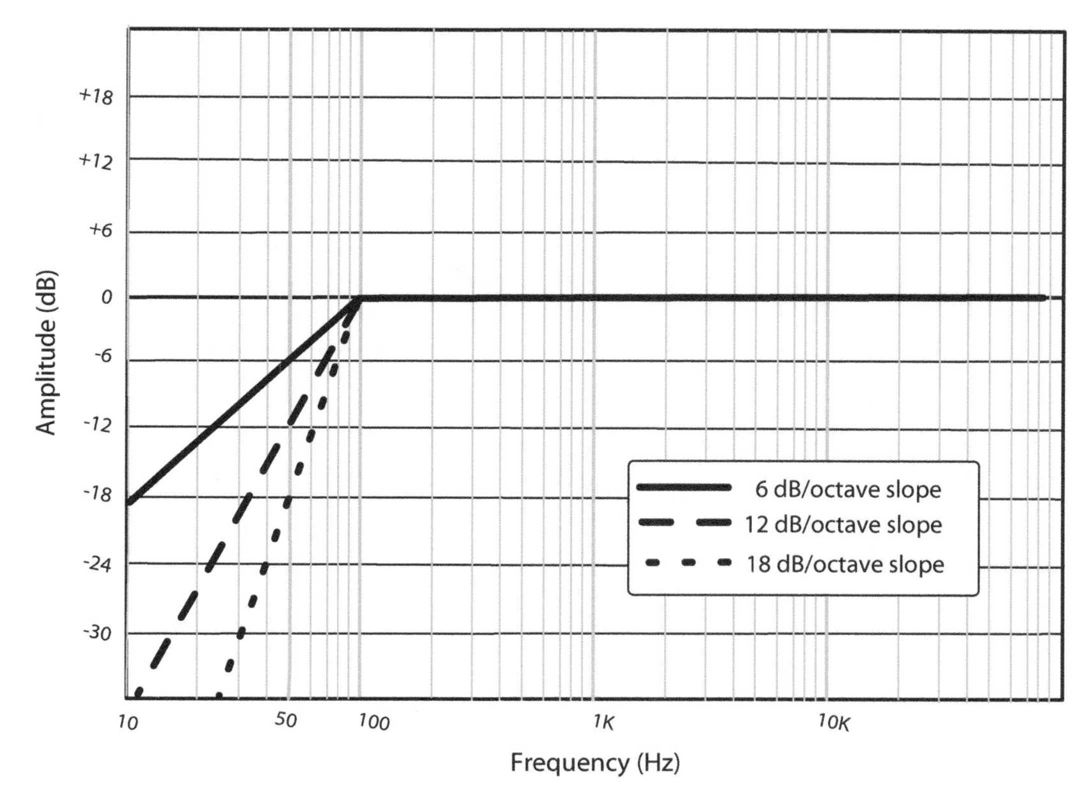

Figure 25.1 Diagram of High Pass Filters with Various Slopes

Compilation albums may require more frequent use of filtering once the levels are set. Pieces of very different timbral makeup might also benefit from longer pause times between the tracks.

High pass filter (HPF): this is a very effective tool for removing low-frequency noise, although it should be used with discretion. The slope of the filter is measured in decibels/octave (see Figure 25.1), and a more gradual slope (e.g. 6 dB/octave) will be less susceptible to ringing. This artifact can be easily heard by applying an HPF with a slope of 18 dB/octave at 100 Hz to active orchestral material, such as louder passages where the timpani are being played. When the play-back is stopped abruptly, a resonance will be heard in the lower midrange that lasts for a fraction of a second. Simply adding a filter at 80 Hz to remove background noise may be too destructive to the musical program, especially in orchestral recordings. A better approach is to filter small sec-tions where the noise is most obvious. For instance, in the case of a truck or airplane passing the recording venue, a filter might be used for those few seconds only, and it can be brought in and out in a subtle way. For persistent low-frequency noise like air handlers in a hall during a live concert, it may be adequate to filter only those moments where the background noise is not masked by the program. Even a rather aggressive filter might be used (100 Hz or higher as needed) during the pauses between movements, or softer passages where the basses and cellos are not playing. Once there is a decent amount of low-frequency information in the music, a filter may not be needed at all, or at least it can be reduced to a lower frequency whenever possible. Low pass filters (LPFs)

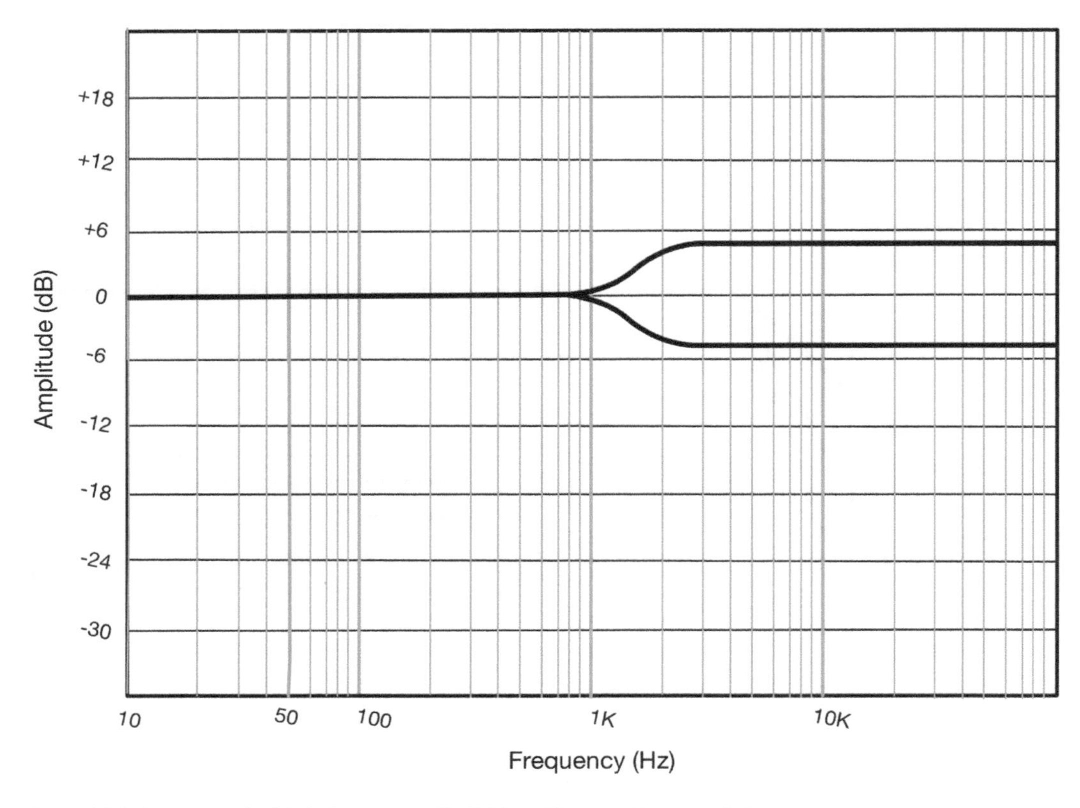

Figure 25.2 Diagram of a High-Frequency Shelf Filter Showing Boost and Cut

have little functionality in classical music production including mastering, as they are designed to limit the range of high-frequency content, which is certainly not desirable.

Shelf filter: the use of a "shelf" rather than a peak/dip filter can produce a timbral shift over a large frequency range with very little boost or cut (see Figure 25.2). Even a 1-dB change can make a significant difference, depending on the filter's frequency. If a high-frequency shelf is set at 6 kHz, the filter will affect the entire frequency range above that point. The opposite will be true for a low-frequency shelf, as all information below the chosen frequency will be affected. Shelf filters can allow for subtle yet effective changes in timbral balance without dramatically altering the overall "character" of a mix.

Peak/dip filter: for changes in the middle range of frequencies, a filter that provides a boost or cut at a specific frequency may be required (see Figure 25.3). The "Q" or bandwidth setting controls the frequency range that is affected by the filter. Lower Q values will affect a wider frequency range and may allow for more moderate values of boost or cut to be used. The end result will be more transparent or "musical" processing of the audio.

25.4 Dynamic Processing in Classical Mastering

As stated in Chapter 24 (Mixing), there is very little need to compress or limit the audio in mastering. Only pieces with extreme dynamic range and contrast may require this type of processing, and

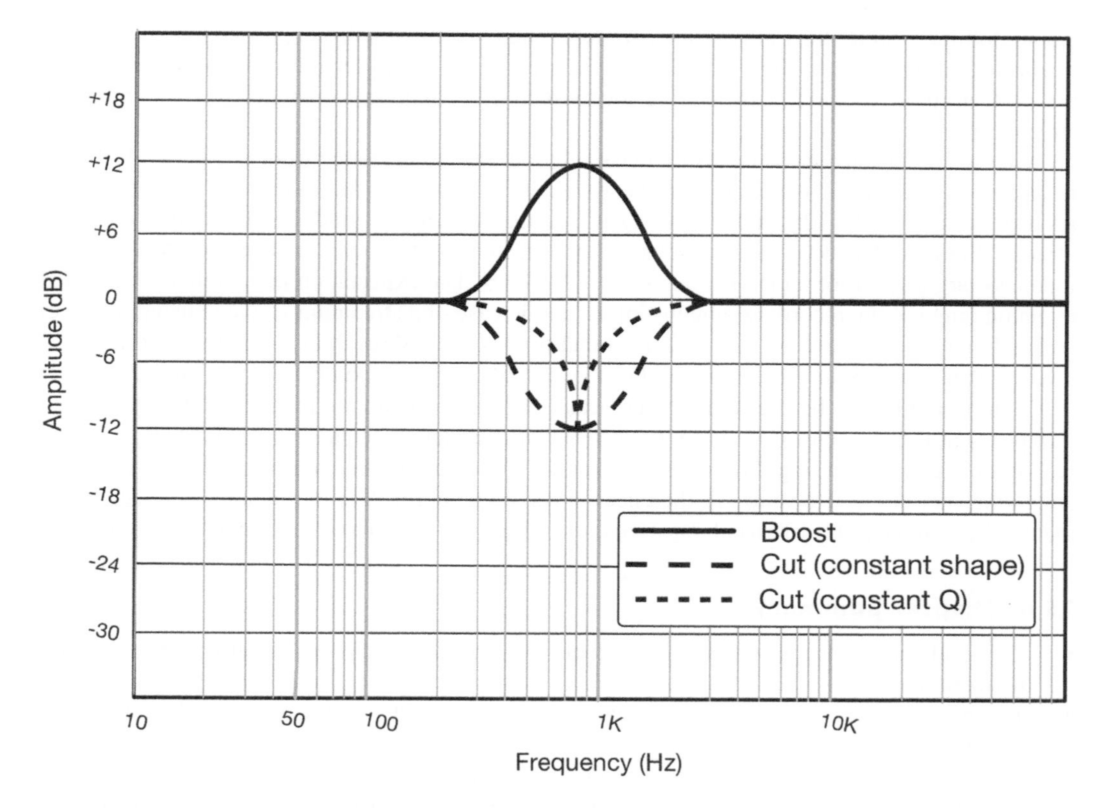

Figure 25.3 Diagram of a Peak/Dip Filter Showing Boost and Cut

it should be implemented with great care and precision. Manually adjusting gains between softer and louder movements or passages should be the first course of action, using dynamic processing only as a last resort. Compressors and limiters with a "look ahead" function are preferred, as they exhibit the most transparency in their operation. This feature allows the processor to preview the audio and apply gain reduction at precisely the same moment as the onset of the transient information (attack) that needs to be reduced in level. This should keep artifacts such as "pumping" to a minimum. Pumping is the sound of a compressor "recovering" or returning to normal level after a momentary gain reduction of the signal. Multi-band compressors do not work well with classical program, where the frequency domain is split into separate bands. The delicate nature of the timbral characteristics in classical recording is easily upset when using this type of processing.

25.5 Final Levels and Loudness

This is a delicate subject, as the various streaming services use different target levels for loudness. Target levels are treated differently as well, whereby most services will lower any content that is at a higher level than a specific target, but not all streaming companies raise the level of content that is under the target. In theory, a lower loudness level means more dynamic contrast, that is, a greater difference between average and peak levels in the program – assuming the program peaks are close to 0 dB full scale (0 dBFS). Loudness is measured in LUFS (loudness units relative to full scale). Most DAWs have the ability to measure the integrated loudness of a track or an entire album "offline", rather than in real time. An integrated measurement, as opposed to a momentary or short-term one, reports the average loudness level of the entire length of the program. This is the most accurate method of measuring loudness and is common to practically all commercial streaming services.

When searching suggested levels for classical program, results range from as high as -13 LUFS to as low as -23 LUFS, depending on what website is accessed. Audio on the Internet is managed per track, but for classical mastering it is better to measure loudness throughout an entire album. This should be done after the relative levels from track to track have been set. A reasonable loudness target level for a classical album is -20 LUFS. The expectation here is that the original dynamic range of the recording remains intact, and in case of a streaming service raising the level and adding soft compression, the artifacts of that process should result in a minimum reduction of quality. Soft compression is a reality and is unavoidable in the world of streaming audio.

Content should also be checked for true peak (dBTP). It is common practice to deliver content that peaks at a maximum of -1 dBFS to ensure that peak levels remain under 0 dBTP, avoiding any audible clipping during the streaming process. When delivering audio for digital release on streaming services, it is best to ask the client or record label for their particular specifications, as many labels are incredibly strict about what they will accept in terms of audio master submissions.

25.6 Chapter Summary

- Classical mastering mostly involves level-matching the tracks and adjusting pause times between pieces.
- The most successful mastering for classical program will be transparent and subtle in its approach.
- If the "secret of recording" in Chapter 6 has been followed (i.e., getting the microphones in the right place), very little equalization, if any, will be needed during the assembly of the master.
- Use of compression should be seen as a last resort, as manual gain changes for managing loud and soft program material can be implemented in a more musical way than any automatic process.

Part VII

Multichannel Recording and Mixing

Chapter 26

How to Listen in Surround Sound

Multichannel recording became a major part of classical music recording in the mid-1990s with the worldwide success of the DVD, which offered video with surround-sound audio. This was followed by the development of the high-resolution Super Audio CD (SACD) for audio only, and later the Blu-ray disc, which allowed for high-definition sound and picture. Support from major record labels waned in the early 2000s, but interest has returned across the entertainment industry with the advent of immersive binaural content. For more discussion of immersive/3D production, see Chapter 28. A few companies have consistently maintained multichannel production over the years, releasing multichannel versions of their catalog along with the standard stereo format.

26.1 Calibrating a 5.1 Playback System

Calibrating the monitoring environment is somewhat more complex for 5.1 listening than it is for stereo. According to the International Telecommunication Union (ITU), the standard layout for five-channel surround sound proposes three front channels spaced at 30° intervals (left, center, right) combined with left and right rear channels located between +/- 110° and 120° (see Figure 26.1). All the loudspeakers should be equidistant from the listener. This setup is common to most consumer home theater installations, preferably using five loudspeakers of the same size and one subwoofer that is used to reproduce the low-frequency effect (LFE) channel. The subwoofer is normally placed near the front channels out of convenience, and it may be off to one side since it is producing only the lowest bass frequencies, which are virtually impossible to localize in a normal-sized listening room.

Ideally, five identical monitors are used, but smaller loudspeakers might be used for the rear channels as long as they are properly calibrated. It is highly recommended that at least the three front-channel loudspeakers be identical; otherwise, the balances will not translate properly between systems. For example, if a small speaker is used for the center channel, a vocal or soloist may be over-balanced in the mix, which will sound too loud or "forward" on another system using matched loudspeakers (i.e., three identical full-range monitors, used for left/center/right playback). Another issue is timbre – even within the same brand of loudspeaker, a smaller model, used for the center channel, may have a different tweeter design or crossover frequency than larger speakers that are used for the left and right channels. Elements in the mix will exhibit a different tonality depending on whether they are routed to phantom center (L/R only) or fully present in the center channel. This difference will be more noticeable when divergence is in play (the process of spreading or sharing a signal across channels in varying amounts), or when a source is automated to move across the front channels. Imagine a vocal that is reproduced at 40 percent center channel and 60 percent phantom center. It will be very difficult to judge the aesthetic result of this placement in the mix, unless all three loudspeakers are matched.

DOI: 10.4324/9781003319429-33

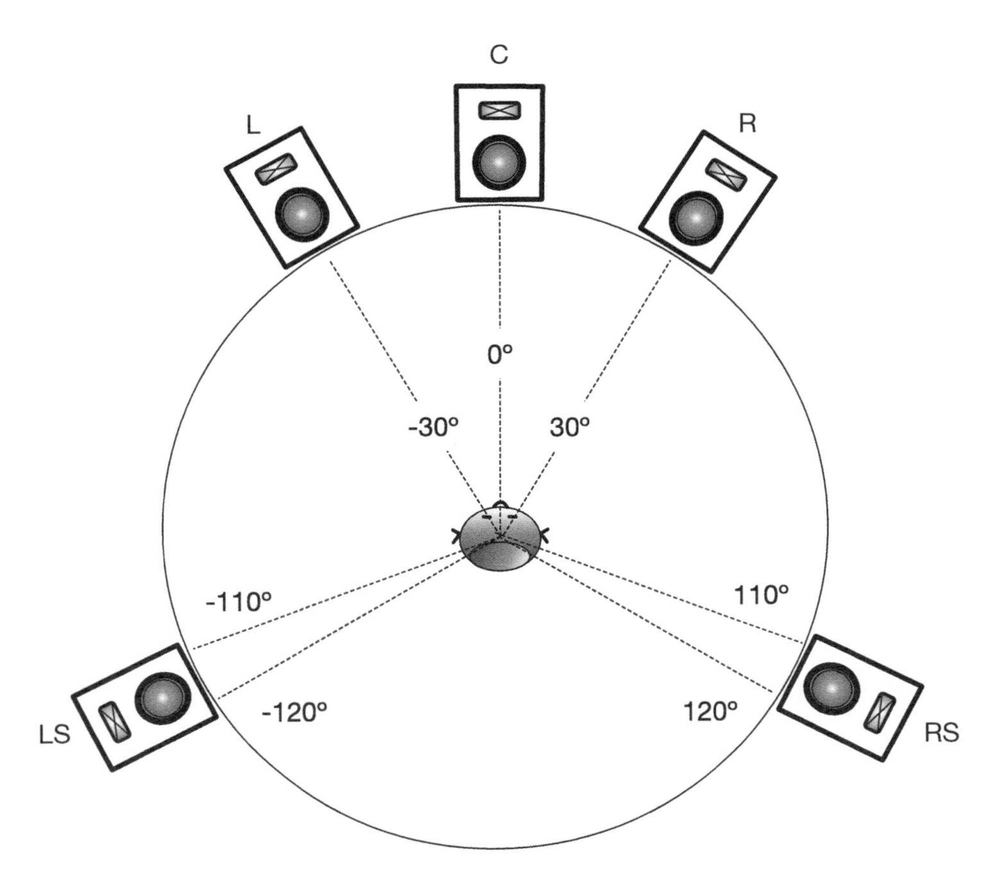

Figure 26.1 Diagram of a Five-Channel Surround-Sound Layout in Accordance with ITU-R BS.775-3 Standard

There are several calibrations and EQ curves used in film and television postproduction, but for music mixing it is generally accepted that all channels be set to the same level and without any playback curve applied. This is accomplished by producing pink noise from each channel separately, and adjusting the output based on the reading of a sound pressure level (or SPL) meter. The output of the subwoofer should be measured "in band" – meaning that within the range of frequencies produced by the subwoofer (normally up to 120 Hz), the level reads the same as for the other channels when measured in the full frequency range. For most surround-sound applications, the standard alignment of the LFE-channel output to the subwoofer is 10 dB above the other channels. The net result is that when listening to the 10-dB boosted output, the mixer will print the LFE channel 10dB lower than it is heard. This is an old film standard that allowed for additional recording headroom on the analog channels of a 70-mm print and that remains in place across the industry [1]. The meter should be positioned at a point that is equidistant from the five main channels and should coincide with the engineer's listening position. Those recording teams fortunate enough to have the option of monitoring in surround at the recording session should allow extra setup time for calibrating the control-room monitors, and there should be a simple method in place for changing the monitoring over to stereo upon request.

Most common SPL meters (including smartphones) use a rather inexpensive omnidirectional microphone. This is fine, since the goal is simply to measure and compare the relative levels of channels. Facing the meter's microphone directly upward for all channels should give an accurate result. The "C weighting" position on the meter will provide the most accurate reading. For checking the alignment between smaller rear-channel loudspeakers and larger, full-range front-channel loudspeakers, the "A weighting" position may be useful. Selecting this option simulates the frequency response of human hearing at moderate signal levels, resulting in a roll-off in the low frequencies. Another possibility is to add a high pass filter (HPF) to the pink noise generator at around 50 Hz, so that the front channels will not reproduce a low-frequency band that is not supported by smaller-sized monitors. In this way, all five channels will be calibrated based mostly on mid-frequency energy.

Polarity should be checked as well, as with passive loudspeakers it is quite easy to mix up the connections, thereby reversing the polarity. A quick way to check is to output the same signal to two loudspeakers at once and listen for a clearly "mono" image between the two loudspeakers. Always using the front-left loudspeaker as a reference, the four other speakers can be compared in turn (left>center, left>right, left>rear left, left>rear right). This should be done after the individual levels have been set.

26.2 Listening in 5.1 Surround Sound

Once calibration is complete, then listening can begin using a few familiar multichannel recordings. The absolute center or middle of the five loudspeakers should be marked on the floor or ceiling, so that the listener can be sure their chair is in the right spot for an accurate impression of the playback. It might be interesting to compare stereo and surround mixes of the same program, if both formats are available, for example, on an SACD hybrid disc. All of the same attributes as in Chapter 2 can be evaluated, with enhanced immersion. Table 2.1 can be used equally well for multichannel listening and analysis, with the added dimension of front-to-back depth. Drawing a picture of what is heard can be very helpful when analyzing the "surround" image.

Through listening through various recordings, with different repertoire and venues, it will become clear that various surround-sound content is created with differing approaches in mind. Moving around the room while listening can be very informative, as front-to-back imaging may shift dramatically depending on the amount of common signal in the five channels. Through the process of evaluating many surround-sound recordings, a preference for style and presentation can be developed that will guide the engineer in their decisions regarding instrument and microphone placement. The following are a few helpful notes for engineers new to surround-sound listening and production.

Center channel: the additional front channel in surround sound is used to anchor center information in place, regardless of the listener's position in the room. The idea came from the film industry, where it is necessary to keep dialog and sound effects associated with action in the center of the screen locked in place. In fact, the evolution of film sound basically transitioned from a mono soundtrack in the middle of the screen almost directly to a three-channel front system, in order to maintain the location of all sound sources relative to their position on screen. Where music mixing is concerned, phantom center imaging used in stereo production can be improved through the use of the center channel, providing additional image stability and helping listeners sitting off to the sides to localize center information in the middle of the listening room. There is a big difference, however, in the sound of a discrete center channel as opposed to a phantom image, because of the crossfeed inherent in stereo loudspeaker listening, More on this in the publication [28.2].

Surround, or rear channels: in combination with the front system, these channels work to "surround" or envelop the listener with ambient sound, or sometimes with direct signals, depending on the production. In a 5.1 playback system, it is virtually impossible to place a mono sound source directly behind the head with any stability. Sound sources that are panned to that position may at times be localized in front of the listener and may have an unpleasant timbral signature, as the brain tries to make sense of the information.

Lateral information, or sound coming from the sides: many practitioners speak of the "phantom image" located on each side of the listener. In fact, there is no way to experience this effect without turning one's head to the side so that both ears can work together in order to experience a center image between two of the loudspeakers – for example, by turning to the right so that the left and right ears are oriented toward the front-right and the rear-right loudspeakers. While facing forward, it is possible to perceive energy from the sides as a combination of signal from front and back, but not as a specific, stable image [2]. In Chapter 28 we look at the addition of side-surround channels, which are specifically implemented to provide stabilized side images and to help bridge between front and rear signals in terms of even envelopment.

26.3 Chapter Summary

- Multichannel systems are an ongoing development; however, there is still demand for 5.1 content.
- Multichannel playback systems require a specific calibration and should be checked periodically.
- Evaluating surround-sound recordings is similar to auditioning stereo content, with the added dimension of front-to-back envelopment.

References

1. Dolby Labs, Inc. (n.d.). "What Is the LFE Channel?" Cited 2023-03-23, available from https://www.dolby.com/uploadedFiles/Assets/US/Doc/Professional/38_LFE.pdf.
2. Ratliffe, P. (1974). "Properties of Hearing Related to Quadrophonic Reproduction." *BBC Research Department Report, RD 1974/38.*

Chapter 27

Recording and Mixing for 5.1 Surround Sound

Techniques and tips are offered in this chapter for successful surround-sound recording and mixing of orchestra and other ensembles, focused on 5.1 presentation (see Chapter 28 for immersive/3D production). Discussions describe stage setups for optimizing a recording for multichannel presentation, and how a recording can be managed for a successful outcome in both stereo and surround sound. The time-saving strategy of generating a stereo and surround mix simultaneously is also presented.

This chapter is entitled "Recording and mixing *for* surround sound", rather than "*in* surround sound", for good reason. For most projects, and certainly location recordings where the focus is on the stereo balance, it may be difficult or even impossible to actually monitor in surround sound at the recording session. Having a multichannel playback system available at the time of recording is effectively a luxury, and trying to implement one may be a major hassle in remote recording situations. A decent room backstage for stereo monitoring is already a rare commodity, while a room that is used in a 5.1 setup will pose additional problems, and probably won't be much help on location if the setup and listening environment are compromised. Also, any guests sitting in the back of the room will have a rather skewed impression of the sound, as they will be seated too close to the surround channels during the recording.

27.1 Surround Presentation Options in Classical Music Recording

The traditional presentation of orchestra in surround sound is to provide direct sound in the front channel- and the reverberation of the hall in the rear channels. Of course, there will be a combination of direct and diffuse sound present in the front, as the main microphones will capture both of these attributes, and are normally assigned to the front channels in a surround sound mix. The rear channels are customarily free of any direct sound sources, and therefore contain only ambient program. This is a conventional setup and makes for an exciting listening experience, as long as the rear channels are active enough and contain an adequate level of "interesting" program (see Figures 27.1 and 27.2).

> A good way to ensure that the surround channels are comprised of "interesting" energy is to position the surround microphones quite near the stage, but facing out into the hall. This will ensure good signal strength overall, adequate dynamic range, and reduced background noise from air conditioning and traffic sound.

DOI: 10.4324/9781003319429-34

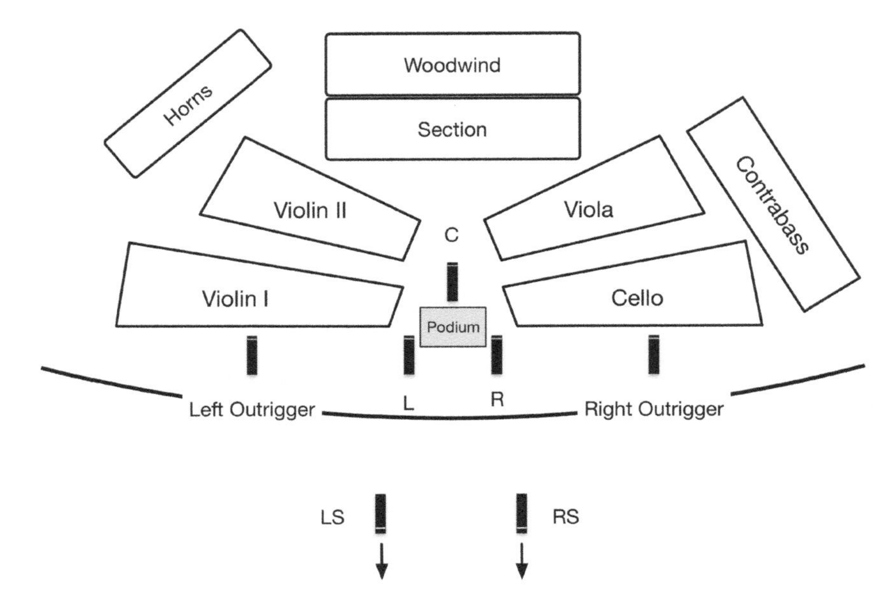

Figure 27.1 Diagram of Orchestra with Multichannel Surround Capture in Place

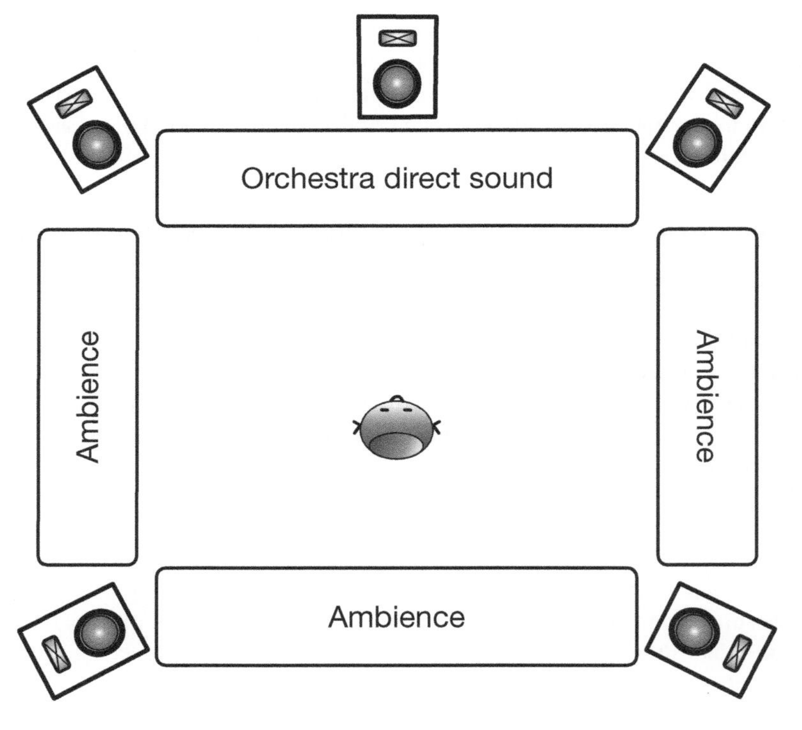

Figure 27.2 Diagram of Multichannel Playback Experience Resulting from Fig. 27.1

Other approaches to surround sound in classical recording involve positioning the instruments and players around the microphone configuration, so that some instruments are localized in the rear channels, or between front and rear. This can be easily accomplished as long as there is enough space on stage, or the effect can be "fabricated" by altering the microphone placement, as in the organ brass ensemble example (in Figures 27.3 and 27.4). Even though the ensemble is positioned roughly in a straight line, the surround microphone position wraps the image of the ensemble around the listener, in an upside-down "U" or horseshoe shape.

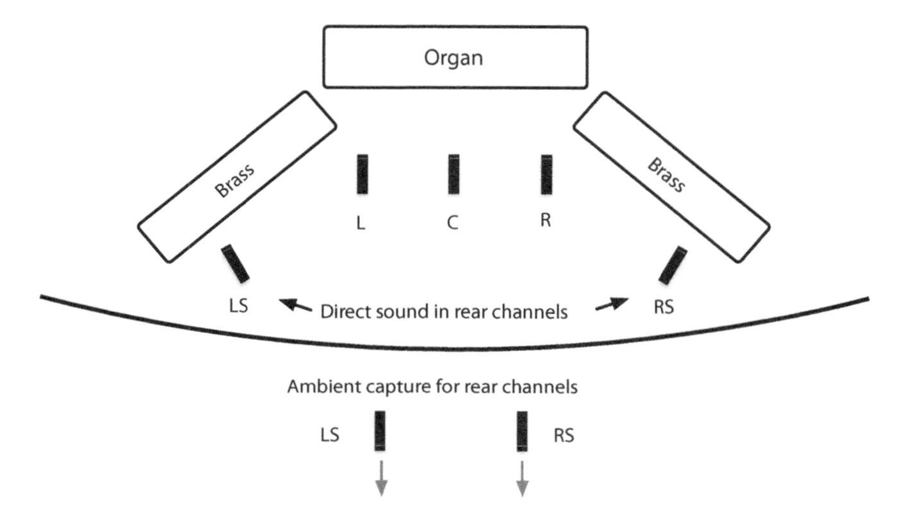

Figure 27.3 Diagram of Brass Ensemble with a Modified Multichannel Surround Capture in Place

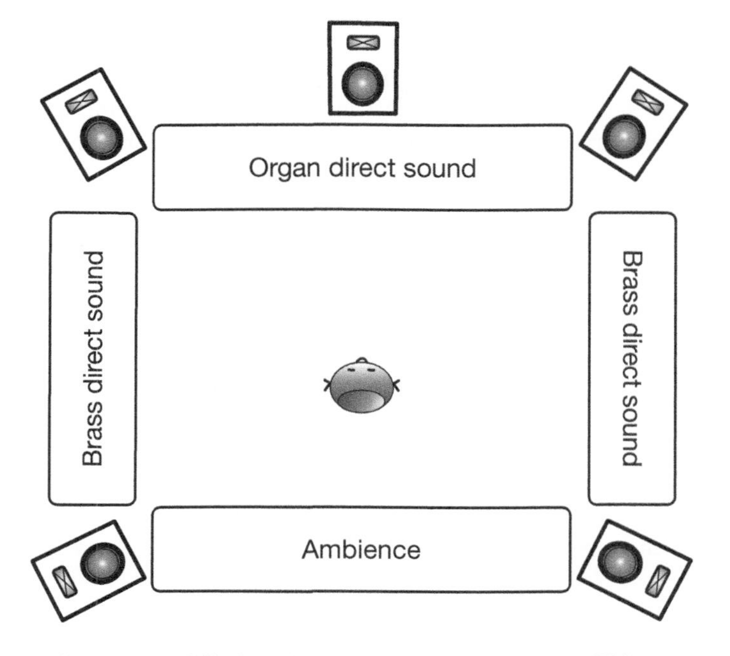

Figure 27.4 Diagram of Multichannel Playback Experience Resulting from Fig. 27.3

This works quite well for chamber groups, whereas an orchestra presented in this way may seem somewhat unorthodox. The traditional stage positioning of an orchestra has been refined over hundreds of years, and the main body of classical compositions have been written taking this into account. The strings are placed in the front and the brass in the back, so they match each other well in level. Once this is undone, it becomes very hard to control the balance after the fact. By pulling apart this standard setup and placing the orchestra "in the round", the natural blend and balance is upset, and the mix will need to be quite artificial in order to make any sense. A few practices might be employed to pull a standard orchestral stage setup out into the surround channels so as to introduce a subtle amount of direct sound out and around the listener.

27.2 Techniques for 5.1 Surround-Sound Recording

While there are many suggested solutions for multichannel recording, commercial surround-sound recordings are executed using many different techniques, usually exclusive to the individual engineer or recording team concerned. In fact, many authors have described configurations that have never been used for commercial recording and have not been subject to any critical feedback, such as a record review in a trade magazine. There is certainly no magic solution based on suspending five microphones over an ensemble and hoping for a decent balance. Refining the placement of a five-channel array takes a great deal of time and patience, trial and error. With the rear channels adding increased diffusion around the listener, there will be even more need to maintain focus and overall clarity in the front channels, which practically mandates the use of spot microphones as part of the mix. This is more often the case for large ensemble recording than with chamber groups.

EXERCISE 27.1 PREPARING A FIVE-CHANNEL ARRAY FOR MULTICHANNEL RECORDING

This exercise describes the installation of a five-microphone array:

1. Add a center microphone to a main system, either on a "tree" stand with the left/right mains, or on a separate stand a few feet in front of the conductor, as in Figure 27.1. The left and right microphones might benefit from a wider spacing, so as to create a hole in the center that might be filled by the center microphone and the center channel. Alternatively, a narrower pickup pattern can be used, and the left and right microphones can be angled to face outward, as is common with the Decca tree. Wide cardioids are an option here, or omnidirectional microphones fitted with an attachment that creates a cardioid polar pattern in the upper-mid and high frequencies (as described in Section 6.1). This feature is built into the design of the Neumann M50 and M150 microphones.
2. Likewise, place a pair of directional microphones – or omni microphones with the previously mentioned attachments – slightly behind the main front microphones, looking out and slightly up into the auditorium. These microphones can be spaced more widely, at around 2 m or 6 ft., and placed at the same height as the main microphones. The distance from the main microphones can be fairly short, around 2 m or 6 ft., because of the de-correlation gained by positioning the capsules so that they are attenuating the direct sound of the ensemble.

3. The microphone positions are then adjusted and refined based on careful listening. The surround pair can be auditioned in the stereo monitors to assess the character of the capture, but a full 5.1 surround playback system will be needed in order to judge how well they work with the main pickup in creating an optimum impression of envelopment. Erring on the close side is preferred for the surround capture, since the signals can always be delayed a few milliseconds in order to increase the perceived distance between the front and rear microphones.

Certain supporting instruments that are placed near the front of the orchestra and out to the sides can be brought partially into the surround field during the mix, for additional immersion. Harp, celeste and piano are instruments that typically come and go in the texture of an orchestral score, so they tend to catch the ear when they enter. This can help to keep the rear channels active and interesting. A second, wider pair of outriggers might be set up and assigned to the back channels to achieve the same effect with the outer string sections, or a closer pair of surround microphones sent to the rear channels will have good effect.

27.3 Use of the LFE Channel in 5.1 Music Mixes

Multichannel music-only mixes such as classical music in surround sound can make use of the low-frequency effect (or LFE) channel and subwoofer playback. The mixer should be careful not to exaggerate the levels since many consumers rely on their subwoofer to provide most of the low frequencies for the entire system. This is certainly the case in most home theater setups, which tend to use five very small "satellite" loudspeakers along with one subwoofer. The LFE channel should be used for low-frequency *extension* rather than *level*, and it should be implemented in a subtle way to avoid any conflict with the main channels. In other words, muting and unmuting the LFE channel shouldn't make the bass content change very much in volume and shouldn't affect clarity in the low-frequency range. Audio in this channel should be filtered a with a linear phase low pass filter (LPF), around 80-115Hz with a gentle roll-off, so there is a minimum of interaction with the low-frequency content in the main channels (also see [28.3]. Orchestral program and organ recordings may benefit from a subtle amount of LFE channel, although the main LCR channels should still carry the full-frequency band. A solo piano album will not benefit from a 5.1 mix over 5.0 (a five-channel mix with no LFE).

27.4 Simultaneous Stereo and Multichannel Recording and Mixing

It is possible to capture for stereo and surround at the same time, with additional channels being recorded along with the usual elements. During the mix process, it should also be possible to balance for both stereo and surround sound, switching between the two mixes while listening. Two separate busses are needed – a two-channel buss for stereo and a six-channel buss for 5.1. This usually results in a higher-quality stereo mix than the process of "folding down" or reducing the 5.1 mix to stereo, where the six channels of the surround mix are assigned to either left or right (or both) channels, at various levels. This "double buss" approach will also allow for separate decisions on panning, allowing for more flexibility. Low-budget projects will benefit from this practice because two mixes can be produced in less than twice the number of postproduction hours, as one general balance should work quite well for both busses. The synchronized mix tracks can then be assembled all at once, so that the track timings and relative levels between the tracks can be adjusted in tandem during the mastering stage. Down-mixing of the surround material can also be a viable option, but the resulting stereo mix should be checked regularly during mixing to ensure quality.

27.5 Artificial Reverberation in Multichannel Mixing

While many five-channel reverbs are readily available, there is a certain amount of added control and flexibility to be gained from combining multiple stereo and mono reverbs. An obvious advantage is that the same reverb can be used for both the stereo mix and the left/right channels of the 5.1 mix. A second reverb can be employed for the rear channels, based on an identical algorithm to the front reverb (for example a "large hall"), but with different settings on certain parameters, such as a lower cutoff for high frequencies or possibly a longer decay time. A third single-reverb channel can be used for the center channel, if necessary.

As mentioned in Section 27.1, the rear channels must be active enough to keep "reminding" the listener that they exist, in order to maintain an appropriate sense of envelopment. A surround capture that is too far from the source will have little dynamic range and may become too familiar to the listener's ear. In this case, as with steady-state signals in general, the "surrounding" effect will seem to disappear until it is muted and unmuted. One simple practice that may help work against this phenomenon is to send signal from the front microphones into the rear reverb, which is then combined with the surround microphone signals. This will provide a more active and interesting source signal, and since the reverb unit reacts to the input in a random way, the rear channels will remain more evident during longer listening periods.

27.6 Practical Considerations in Multichannel Mixing

With the additional loudspeakers in play, it is advantageous to place the monitor screen of the playback computer off to one side. This way, with the front three loudspeakers being the same size, there is no risk of the video monitor blocking the center loudspeaker, or vice versa. Another option is to use two screens and place them either side of the center loudspeaker. When mixing to picture, it is nice to have a large-format video monitor above the center channel for picture playback of film or concert footage.

> Mixing in surround will be problematic for a producer/engineer team – if one sits behind the other, they will have very different impressions of the front-to-back balance. Sitting side by side may be the best compromise, while taking turns checking the balance and overall presentation from the center position.

Documentation is very important in multichannel deliveries. There are several standards for channel order, so it is best to include the channel name in each sound file, and the files should be numbered so that they sort in a specific order, rather than alphabetically. Many engineers and archivists insist on including the creation date in the filename, among other details. Asking the client for their file-naming convention and preferred channel order is a good idea. The final filenames might look something like this, as an example of how to structure the naming of a combination stereo and surround delivery:

01_StraussOpus86_frontleft.wav
02_StraussOpus86_frontright.wav
03_StraussOpus86_surrleft.wav
04_StraussOpus86_surrright.wav
05_StraussOpus86_center.wav
06_StraussOpus86_lfe.wav
07_StraussOpus86_stereoleft.wav
08_StraussOpus86_stereoright.wav

Stereo mixes may also be delivered in a separate folder, as an interleaved file (rather than separate left- and right-channel files). In newer terminology, the more common name for the stereo mix is the "2.0 mix".

27.7 Chapter Summary

- Listening and working in surround sound requires that the engineer refines and expands their "peripheral" hearing, no longer focusing on sound coming only from the front of the listening room.
- Mixing for stereo and surround formats concurrently can save time and benefit low-budget projects.
- The subsequent chapter (Chapter 28) covers immersive/3D production.

Chapter 28

Recording and Mixing for Immersive/3D Content

This chapter expands on traditional 5.1 surround-sound methodologies and includes a discussion of the ins and outs of delivering immersive content to a client or record label for distribution via music-streaming services.

Immersive music production is now very popular with artists and record labels, as they seek to provide consumers with a more engaging listening experience. Currently, the priority is binaural reproduction of immersive content over headphones, due to its widespread availability and cost-effective delivery to the consumer. Listening to stereo program over headphones is fairly limited in terms of its presentation of soundscape (see Section 2.6), although some listeners enjoy the intimacy of hearing music localized within the head. Immersive mixes using binaural encoding allow the music to be externalized – in front of, above, around and behind the listener – which provides a more realistic experience. A more in-depth explanation of binaural encoding can be found here [1].

28.1 Setting Up an Immersive Playback System

Height channels are becoming more common in film and music production and are being introduced into broadcast formats in Asia and Europe. Playback channels are now expanding upwards to 9.1, 10.2 (using two subwoofers) and the Japan broadcaster NHK's 22.2 format. A simple iteration of surround plus height is known as 9.1. This is a combination of the conventional layout of five loudspeakers at ear height plus four additional monitors positioned over the front and rear left/right loudspeaker pairs, along with a subwoofer for LFE reproduction. Although exact positioning may vary, these height channels create a second ring above the listener, providing elevation and adding a third dimension to the sound (see Figure 28.1).

Dolby Atmos is an immersive system that utilizes a "bed" of 10 channels, working in combination with audio objects. These objects are positioned using metadata in a proprietary file that contains the audio and all positioning information. Placement of the objects is then controlled locally by the playback renderer in each listening room, specific to the available loudspeaker configuration. Sony 360 Reality Audio (RA) can accommodate many channels, including a lower LCR frontal staging, and also incorporates object-based panning technology. While the total number of channels is variable and can currently reach an upper range of 64–128, the recommended playback configuration with a *minimum* channel count for effective immersive audio reproduction across most platforms is 7.1.4. This is based on a traditional 5.1 system with an added pair of side-surround channels at +/-90 degrees to the listener, plus four height channel loudspeakers, as pictured

DOI: 10.4324/9781003319429-35

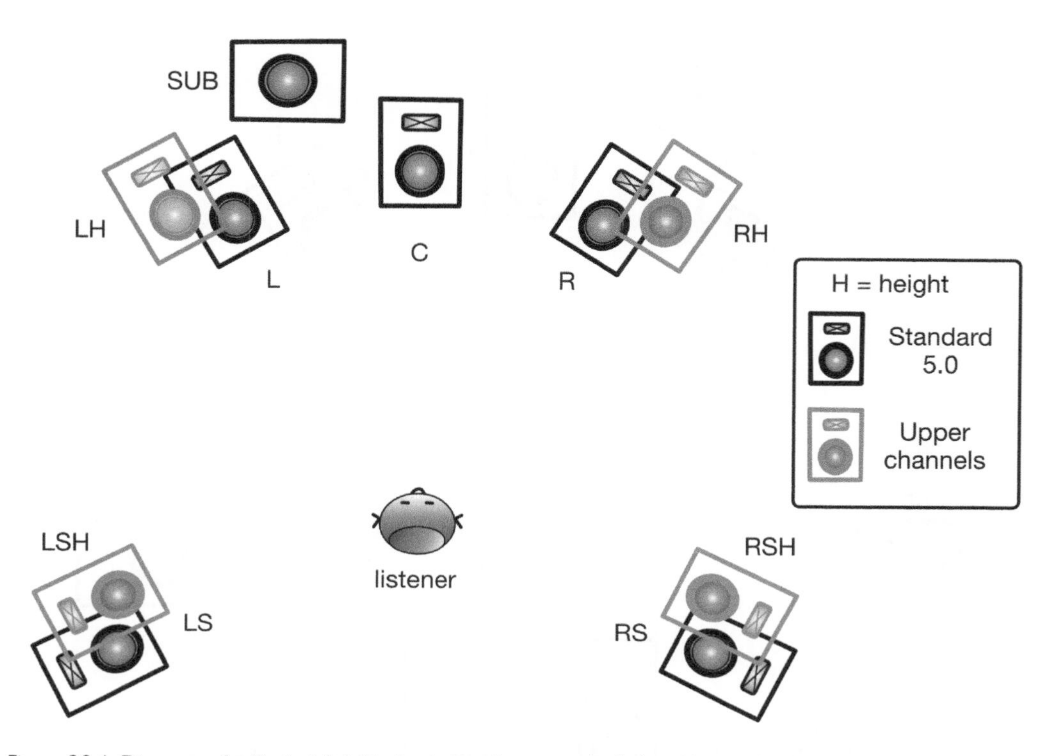

Figure 28.1 Diagram of a Typical 9.1 Playback Configuration Including Height Channels

in Figure 28.2. Those who wish to read further can investigate the standards for multichannel loudspeaker placements, ITU-R BS.2051-0 [2].

28.2 Immersive Recording and the Use of Height Channels

Since the film industry introduced height channels into movie-cinema playback systems, the concept of elevation in surround sound is starting to become a reality in commercial recording. Engineers interested in being involved in the future of multichannel recording and mixing should be actively experimenting with height channel production. As discussed in the previous section (Section 28.1), a number of suggested configurations exist for the number of channels and for loudspeaker placement, but the field is still emerging. The best course of action is to develop and evaluate various techniques that inform the engineer on how they may use height channels to their advantage in any immersive format.

Placing height microphones above the main capture and the traditional surround capture pair may be a good starting point, with the capsules facing away from the source (upward, or back and up). Typically, 1 m or 3 ft. of vertical spacing is adequate, and coincident techniques may also provide a decent immersive field. Different polar patterns may offer certain advantages over others, and the orientation of the capsules will dictate the characteristic of the capture. The main objective is to realize a height content capture that is mostly de-correlated from the main system, so that the image of the ensemble or source is not "lifted" as the height channels are introduced into the mix.

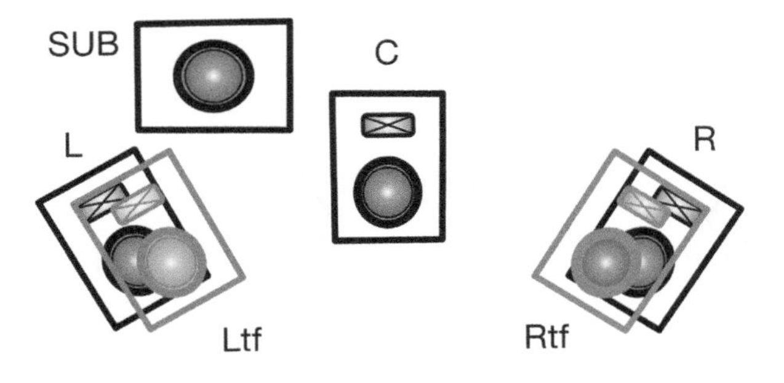

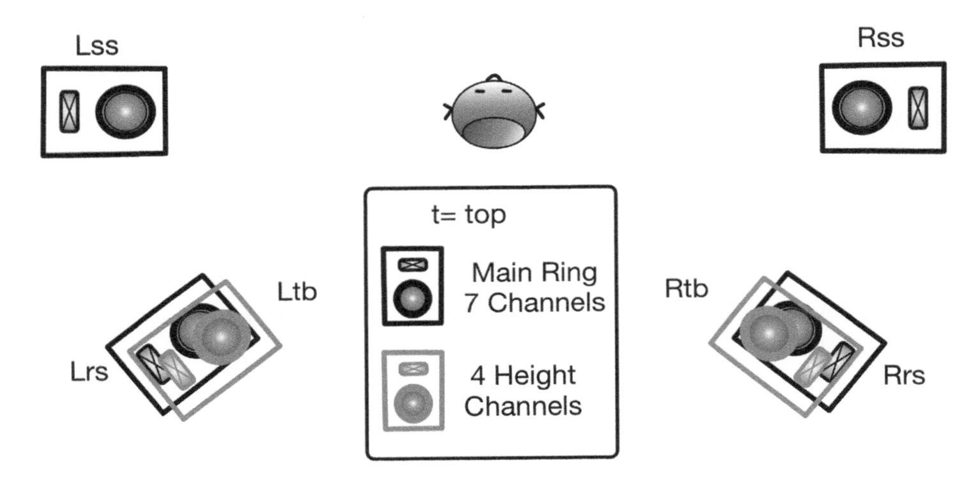

Figure 28.2 Diagram of a 7.1.4 Playback System for Immersive Music Mixing

EXERCISE 28.1 PREPARING HEIGHT MICROPHONES FOR AN IMMERSIVE RECORDING

1. Start with a main LCR, side- and rear-surround microphone setup (seven-channel array) for the main channels of an immersive mix.
2. Add height channel microphones above the main L/R and above the rear-surround pickup, with the capsules looking away from the source (up and forward, or back into the hall).
3. A cardioid polar pattern is a good starting point, although bidirectional capsules may be an interesting choice, while omnidirectional capsules with pressure equalizer (sphere) attachments will provide good decorrelation above 1 kHz (as discussed under "Decca tree" in Section 6.1, Main Microphone Systems).

4. In a room with a properly calibrated immersive playback system, open the LCR and other surround signals with the associated routing to the main ring of loudspeakers, then introduce the four height signals into the upper playback channels and evaluate the result.
5. As more gain is added to the height channels, a small amount of delay may be required, possibly 1 or 2 ms, to keep the source image anchored at the ear height of the listener.

Artificial reverberation can be used in the height channels on its own to create a certain impression of immersion, but the live capture of height information in the recording venue will provide the most realistic source. A combination of live signals and the support of artificial reverberation may be the best solution, depending on the circumstances.

28.3 Immersive Mixing

Mixing immersive music content has become a routine undertaking for many audio professionals, as artists and labels are seeking to provide 3D content to their audiences. Binaural processing or rendering results in a spatially encoded two-channel audio file that can be enjoyed on any decent pair of headphones. The processing aims to externalize the image outside of the head in the same way it would be heard on an immersive loudspeaker system. It is now possible to create immersive mixes using headphones only, without the need for a large loudspeaker setup. Binaural renderers still have a long way to go and are not always accurate or effective. More reading is available here regarding personalized ear and head measurements or head-related transfer function (HRTF) measurements and their benefit in binaural reproduction [1]. Use of center channel and side/rear surround channels follow practices discussed in Chapter 27, although the side surround channels open up new possibilities in spatial stability, as they help to fill the gap between front and rear channels. They might also be used effectively to bring sources out of the front channels. For a deeper discussion on the use of center channel vs. left/right phantom center, please refer to [3].

28.4 Immersive Delivery

These days, immersive music content is available from streaming services such as Amazon, Apple, Deezer, Tidal and many others. The most common format is a binaural rendering that is optimized for headphone playback. Almost every service is adopting a different method for streaming 3D content, from different binaural processes and profiles to varying "interpretations" of spatialization settings in the metadata, as in the case of a Dolby Atmos (ADM) file. Sony 360 RA uses a dramatically different profile than Dolby, while Apple creates its own rendering based on the multichannel content of the Dolby file.

All this is to say that for immersive mixers and mastering engineers, delivery is a moving target. There is no consistent way to know how an immersive mix will sound over headphones, now or in three years from now, as profiles and filters change over time. Checking as many different platforms as possible will help inform you on the overall outcome, but the best practice is to create and refine mixes while monitoring on an immersive loudspeaker setup and to hope for the best for all the binaural versions that are generated thereafter. One important note is that LFE filtering is not consistent over various streaming services, so it is best to avoid heavy use of the LFE channel. Sony 360 RA does not support an LFE channel, whereas other platforms sum the LFE with the main LCR as part of the binaural processing. In some cases, there has been evidence of low-frequency cancellation during the summing process when aggressive amounts of LFE are used (see [4] for more detailed discussion of LFE use in immersive mixing). The advice here is to make sure the main channels contain a full-range signal including the entire low-frequency range, so the LFE is not needed or its use can be restricted to subtle LF "extension".

28.5 Chapter Summary

- Immersive recording and mixing introduce the parameter of elevation into surround-sound production through the use of height channels.
- Although there are many formats with different channel counts, a minimal setup for loudspeaker playback for music mixing is 7.1.4, which includes four height channel loudspeakers.
- Immersive audio delivery is currently hit and miss, as different streaming services and immersive platforms are processing the multichannel files in different ways for binaural reproduction.

References

1. Roginska, A. (2018). "Binaural Audio Through Headphones". In A. Roginska and P. Geluso (Eds.), *Immersive Sound: The Art and Science of Binaural and Multichannel Audio*, 88–117. New York: Routledge.
2. Recommendation ITU-R BS.2051-0, Advanced Sound System for Programme Production, accessed July 15, 2023. https://www.itu.int/dms_pubrec/itu-r/rec/bs/R-REC-BS.2051-3-202205-I!!PDF-E.pdf.
3. King, R., Theriault, M., and Massenburg, G. (2023). "A Practical Approach to the Use of Center Channel in Immersive Music Production". Proc. of the 155th Convention of the Audio Eng. Soc., New York, October 26th.
4. King, R., and Theriault, M. (2023). "LFE, Friend or Foe? The Pros and Cons of the Low Frequency Effect Channel". Proceedings of the 154th Convention of the Audio Eng. Soc., Helsinki, May 13th.

Chapter 29

Film Score Recording and Mixing

An entire book could be written on recording film scores, as there is such a vast amount of detail and so many variables to discuss on the subject. This chapter will provide a full overview of the process, including logistics, microphone positions, DAW preparation and a breakdown of all personnel involved. While the basic principles of classical music recording still apply to orchestral film score recording, many adjustments and adaptations are required in order to achieve certain specific aesthetic requirements of the discipline. The musical score resides in and among various sound effects and ambiences and will always be overshadowed by dialog, which is the most important element in the soundtrack.

29.1 Film Scoring Personnel

Director: Hollywood-style feature films incorporate a large crew of production people, with the most significant of these being the director. They will normally be a part of all facets of the filmmaking process, and in postproduction they will focus on the picture cut as well as all aspects of sound; this role includes attending the music-scoring session. Most directors lack any formal musical training, and as such they tend to take a more simplistic approach to the music score than would a musician or music producer. Rather than evaluating the actual music, they often focus on how the music affects them as they watch the film. Composers may at times receive rather insulting or simplistic criticism of their work because the director lacks the vocabulary needed to properly discuss the score in musical terms. The picture editor is normally the "right-hand person" of the director during postproduction, simply because of the amount of time they have spent working on the film and with the director. This means that the director may frequently take their advice over that of the members of the music team, including the composer and the music editor. No one should be surprised if they hear the picture editor making musical suggestions during the orchestra recording sessions, as this is a normal occurrence.

Sound department: sound and music are normally separate departments on a feature film, since the two tasks are both quite involved and remain, in many ways, unrelated until the final mix (also called rerecording, or the "dub"). This is the point at which dialog, music and effects are combined. The sound department includes the sound supervisor, dialog editors and sound designers, rerecording or dub mixers and the music editor.

Music department: frequently run by the composer, the music department includes the music supervisor, the composer and orchestrators/arrangers, the musicians and the contractor, as well as the scoring engineer and the assistant/DAW operator. Lower-budget feature films offer the entire music budget to the composer, and in turn the composer pays all charges, including those for all

DOI: 10.4324/9781003319429-36

personnel and for recording and mixing studios. Their resulting fee is the remaining balance at the end of the project.

29.2 The Role of Music Editor

Music editors normally fall under the remit of the sound department, mainly because at the time of hiring, the music department may not have been formed. A music editor is normally hired well before the composer, and they may be contracted for as long as eight months to work on a single film. They come in early in the process and sit with the director to watch a rough cut of the film without music, to decide where the music cues will appear in the film. This is called *spotting* the picture. Some directors have a very strong idea about this from the beginning, while others will rely on input from the music editor when deciding where music cues might help the action and the storytelling throughout the film. While poignant music can help influence the emotional state of an audience, music cues are also used as a transitional effect, providing a bridge between scenes. Additionally, the score might be used as a sonic backdrop for a montage, which is often used to give the audience a break in the action or the progression of the story.

Once the music "starts" or cues have been confirmed, along with their desired length, the music editor goes about the task of finding "temp" track music for each start. The temp track is used as a placeholder, providing the director with a sense of how music will support a scene. The editor will look for music that fits with the director's description of the scene and of the synopsis in general.

Music editors will have a large library of music on hand, categorized by style, theme, ensemble type (orchestral or other) and composer in the case of existing film soundtracks, if that is an efficient way of quickly finding temp music that might fit a scene. The goal is to satisfy the director in terms of their general concept of how they envision music helping to support their film. Even though the temp track will be completely replaced with newly composed music, the temp audio still needs to convey the right emotion or message as the picture cut is being refined and the director and producers are evaluating the early stages of postproduction. The following section outlines an example timeline of how music might be managed during a film production.

29.3 Film Scoring Timeline

PHASE I

1. The music editor meets with the director to spot the film.
2. A list of music cues or "starts" is compiled, including source audio and actual score.
3. Naming is applied to each cue – 1M1,1M2, 2M1, 5M5, and so on, whereby the first number is the reel of film the cue appears in, "M" stands for "music" and the second number applies to the order of cues within each reel.
4. The music editor researches and places temp music for each cue, and then presents the temp score to the director for their approval.

> "Source" or "source audio" refers to music that is actually part of a scene, such as a song coming out of a car radio, or perhaps, in the case of a nightclub scene, the sound of a jazz trio playing live in the background.

PHASE II

1. A composer is hired and brought in to meet the director and watch the film with the temp music in place.
2. The director communicates their vision regarding the music soundtrack.
3. The composer writes new original music and makes a MIDI demo or mock-up for each cue. 4. The composer presents the demo to the director and then makes changes based on their comments.
4. The scoring engineer is hired, usually requested by the composer.
5. The recording sessions are scheduled; the number of mix stems is confirmed with the scoring mixer. This is to specify the number of outputs needed for the music "rig" (DAW) at the final mix stage, as each department (dialog, music and sound) will play back their tracks on separate systems.
6. The music editor prepares each cue in a separate DAW session, importing picture, MIDI demo, rough edits of dialog and sound design and the click track for each cue.
7. The assistant engineer/DAW operator coordinates with the music editor for DAW setup and imports the scoring engineer's microphone input template into each cue.

PHASE III

1. A contractor is hired; they book all the musicians and attend the recording sessions.
2. All the music is recorded; the music editor edits/cleans up each cue, working with the composer.
3. The scoring engineer mixes each cue out to stems (i.e., separate mixes of strings, brass, percussion or synths – basically any musical elements that were recorded separately).
4. The music editor or assistant engineer imports the mix stems into new DAW sessions, one for each reel of film.
5. The music editor attends the "dub" or final mix, representing the composer and, to some extent, the scoring engineer. The final mix is where the dialog, music and sound effects (FX) are combined into a composite soundtrack for the film.

> Various DAWs can be used effectively in film score recording, but the most pervasive is Avid's Pro Tools. This is because of its efficient system of organizing files into playlists and folders as well as its other functions, including tempo-map/click-track generation and timecode management, not to mention its ability to handle MIDI and audio files together.

29.4 Recording Orchestra for Film Scores

It probably goes without saying that the Decca tree or some iteration of it will be used for the main microphone system in film scoring. Use of the center channel in film soundtracks is absolutely critical across all sound departments. Dialog of course is almost always placed in the center channel, except for the occasional line spoken by a character off-screen or far off to one side of a picture shot. Ambiences in sound design take advantage of all channels to some degree, and special effects need to be anchored to any onscreen action, including the center of the screen. Music needs to be evenly presented across the entire cinema, and the center channel can help greatly in that respect. For viewers sitting off to the sides, the center channel will anchor the image so that the music score doesn't collapse to one side of the room. Adding an LCR system over the woodwind section can help as well, for the same reasons. This system can be supplemented with additional inputs per instrument as needed.

In terms of recording perspective, it must be taken into account that the music score will be competing with dialog, ambiences and sound effects. A closer capture will be necessary than would be normal in traditional orchestral recording. A Decca tree using microphones that are more directional than traditional "omnis" is a good starting point (see Section 6.1). Microphones with a sphere attachment behind the capsule will have more "reach", as they become more directional in sensitivity into the higher frequencies. The original Decca tree Neumann M50 becomes narrower in polar pattern starting at 1 kHz and reaches a true cardioid shape at around 4 kHz. There are other, less costly options for achieving this response, such as the Schoeps MK2H with KA 40 attachments, the DPA 4006 with APE40 spheres or Neumann's KK183 with KM body. See Figure 29.1 for an example.

The placement of other microphones may not need to feature a closer proximity to each instrument; traditional placements will work well, although more microphones may be needed, and they may be introduced into the mix at a higher level than usual, in order to properly present the characteristic sound and balance of a film soundtrack. Some isolation between sections may help in adding an additional level of control, such as gobos separating woodwinds and brass, percussion and so on. Typically, the net result is a clear and detailed direct signal that can compete with other

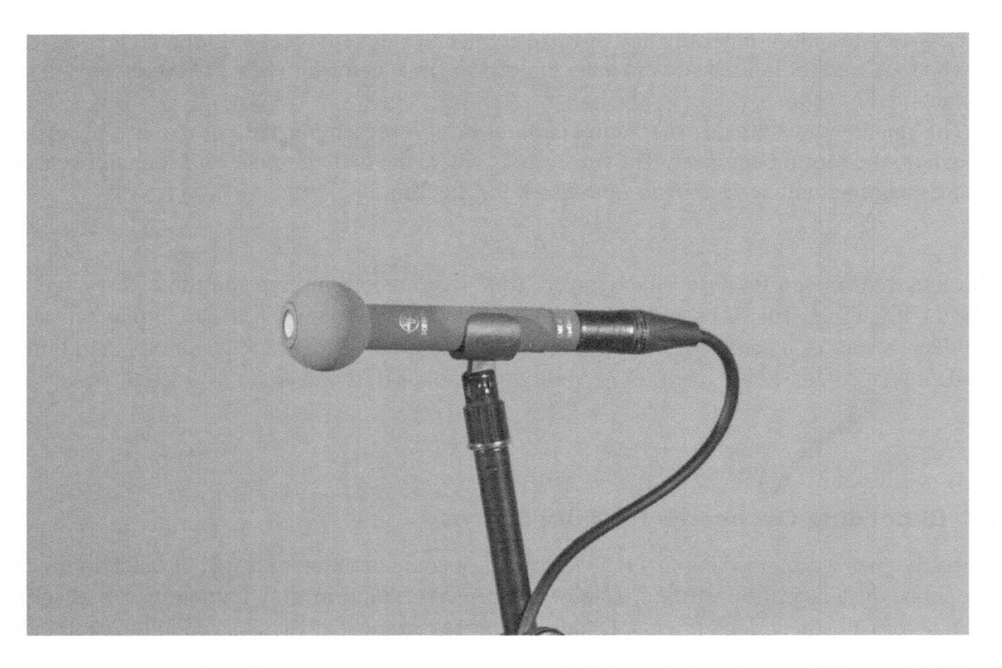

Figure 29.1 Example of an Omnidirectional Microphone with Spherical Attachment (Schoeps MK2H with KA40)

Credit: Jack Kelly

sound sources in the film, along with a good amount of reverb running for general support and sustain. Film scores are usually recorded in large, mostly dry studios ("scoring stages"); however, a good hall can work very well if the microphone setup is appropriate.

As with any immersive/surround-sound recording, it is worth adding height microphones above the Decca tree and a pair farther back (normally above the rear surrounds) as well as side- and rear-surround channels. More and more film scores are being mixed for compatibility with Dolby Atmos playback, so it is a good idea to plan for an immersive capture as part of the microphone setup. More detailed explanation of immersive recording and mixing can be found in Chapter 28.

29.5 Recording Other Ensembles for Film Scores

Cinema sound is almost always presented as a "larger-than-life" experience. This helps in transporting the audience to another time and/or place as part of the storytelling effort. Solo instruments and small ensembles might be presented in an expanded way, in that the "size" of the instrument is exaggerated to match the giant size of the screen. String quartet will work well under a Decca tree with a narrow spacing, placed at a lower-than-normal height. Solo instruments such as piano or acoustic guitar will benefit greatly from an LCR capture, so that the source can be presented with stable imaging and spread across the three front channels. For any of these variations, surround and height microphones may be added for a more realistic immersive presentation, especially when the instrument has a featured role in the soundtrack. For instance, if piano is a major part of the score and appears at important moments throughout the film, it makes sense to record all the piano cues with a fully immersive capture.

When multiple instruments are recorded separately for the same cue, there will be less need for a multichannel capture of each instrument in the score. Multichannel reverberators are an efficient alternative way to create an immersive field, and in some cases, may allow for a smoother and more rapid recording session if the microphone array can be kept to a simple three-channel LCR pickup. For piano, a wide LCR out in front of the instrument works well, and the L and C microphone positions can be optimized for AB or L/R placement for the stereo mix. This will avoid the need for a fold-down of the three widely spaced microphones (see Figure 29.2).

For an instrument like the guitar, a small LCR setup will be useful. This might be realized with three matched microphones, or may be presented using a mono pickup in conjunction with a stereo pair of microphones of a different type, such as a ribbon in the center with a stereo pair of condenser microphones for the L/R capture (see Figure 29.3). Using different microphone types provides for options in timbre blending, depending on which microphone or microphones are dominant in the balance. Either way, the three microphones should be carefully placed at equal distance to the source so that they are more-or-less phase coherent. This will allow them to be mixed at equal gain without the risk of comb filtering during the stereo fold-down process.

29.6 Scoring Session Preparation

The DAW sessions for each cue must be prepared and checked well before the actual scoring sessions. Time-code starts, a click track with the right amount of count-off, and bars and beats all need to be lined up. Time code ensures that the cue start lines up exactly with the picture. Click track and count-off enables the orchestra to start the cue at precisely the right moment in the film, and the bars and beats readout in the DAW allows for easy navigation throughout the cue in sync with the printed score. A separate DAW session needs to be prepared for each music cue in the film because of the bars and beats functionality, as each music cue needs to start at measure 1 to correspond to

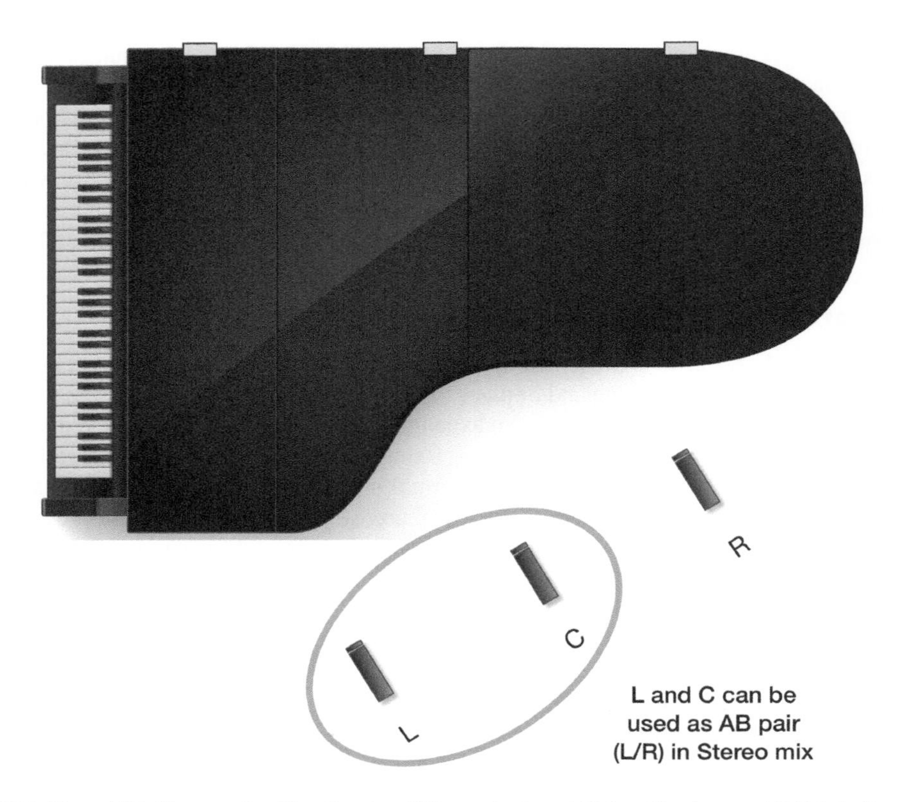

Figure 29.2 Piano LCR Capture for Film Scoring, Where the L and C Are Used as an AB Pair in Stereo

Figure 29.3 Three-Microphone System for Recording Acoustic Guitar for Film

the printed score. Playlist organization will also be specific to each cue. The following exercise navigates through the various steps in preparing a cue for recording. This is normally prepared by the music editor on a feature film, but for a low-budget project the engineer may have to manage all this work themselves. In this example, Pro Tools is used as the DAW, because of its widespread presence in film score production.

EXERCISE 29.1 PREPARING THE DAW: PICTURE, PRERECORDS, CLICK, TEMPO MAP

1. Ask for confirmation of sample rate and timecode frame rate (consult the composer or arranger, or the music editor), and set the DAW to those values (for example, 48 kHz and 24 frames).
2. Set the main counter to Bars and Beats and the sub counter to time code (TC).
3. *Import movie*: picture will be split into separate reels, beginning at reel 1. Spot it to the same timecode as the burn-in shown in the video. Also import any audio that comes with the video, along with any guide tracks (dialog/FX/temp score, etc.). Group the video and its audio tracks together so they stay in sync. If the burn-in TC starts at 1 hour, place the film at 01:00:00:00, or the picture start with an 8-sec countdown and a "two-beep" at 00:59:52:00, so that the reel begins at 1 hour after that.
4. *Import audio*: spot any prerecorded files to the TC number normally listed in the file name of each audio file, make a group for all the prerecords. If you have a TC start in the filename, you can use the "import audio to selection" feature by placing the cursor at the start time of the files before opening the import audio menu.
5. Import the printed click track (if the audio file is available). This will have an earlier start time than the prerecord files because of the count-off.
6. Import the tempo map from a MIDI file (*Import MIDI*): a .mid file will be provided by the composer, as a MIDI export from their Sibelius or other music-creation software. It is not always necessary to import the actual MIDI files; these will not be needed during the scoring sessions if the composer has provided audio files created from their MIDI mock-up.
7. Ensure that the tempo map lines up all files, including the click track. The first bar of music should be aligned with the prerecords so that measures in the printed score line up with measures in the DAW throughout the cue. The standard for feature films is to start the music in measure three, so that a two-bar count-off occurs in measures one and two. In this case, measure one should be in line with the beginning of the printed click track.
8. In Pro Tools, the tempo map can be shifted using the Event tab, then Time Operations/Move Song Start. Place the cursor in the correct start position before opening the Time Operations window; check the appropriate boxes and radio buttons, then hit Apply (see Figure 29.4).
9. Referring to both the printed music score (or a PDF) and the audio files, make sure the click track, prerecords and tempo map all line up. Check that measure numbers in the score line up with the bars and beats of the tempo map, that the picture timecode burn-in matches the timecode display in Pro Tools and that the picture is running in sync. The composer or music editor may need to confirm this last issue, although a list of TC starts for each cue should be available. Some composers will indicate timecode start times in

the score, which is even more helpful. Also check that the click track counts off properly into the beginning of the music – typically, there are two bars of click before the music begins.

10. Move the video to the second screen (you will need to use two screens for film scoring), adjusting the picture to fill the screen by right-clicking on the image.
11. Open the Big Counter and set it for Bars and Beats; then move it to the bottom of the second screen under the picture.

Figure 29.4 Time Operations/Move Song Start Function in Pro Tools

Credit: Image courtesy of Avid Technology, Inc.

29.7 Notes on Session Preparation

- Picture (video files) should always be saved to and played back from a different hard drive than the audio drive used for recording. This is to avoid any hiccups or glitches in picture playback while recording (very important when the director is attending the sessions).
- In Pro Tools, if the frame rate on the left side of the video track is displayed in red type, the frame rate of the video does not match the session frame rate. This should be corrected.
- Some films are organized so each reel starts at a different hour, even though each reel is only around 15 minutes in length. For example, Reel 3 would begin at 03:00:00:00, with an eight-second countdown, so the actual start time would be 02:59:52:00.

- The "two-beep" is literally the beep we hear at the two-second mark during the countdown (two seconds before the picture starts). This can help set the picture to the right timecode position when there is no TC burn-in.
- "Prerecords" normally include the composer's MIDI mock-up (demo) along with any elements that were recorded before the scoring session. The demo is usually separated into stems so that the balance between prerecorded elements can be adjusted easily, or certain elements can be muted while they are being replaced during the scoring session.
- It is recommended that a printed click track be used rather than the DAW's click generator, for several reasons. A printed click can be muted to end exactly on the beginning of the last note of a cue (or earlier), allowing for a nice clean ending without the click bleeding into the room from all the headphones and into the microphones. Also, a printed click can be automated in level, so that it is softer when not needed much, and louder when musicians need to hear it more. Thirdly, although less of an issue these days, if a computer is running out of CPU power or DSP in general there is a risk that the first click may not "fire" at times, because of a lag in the playback buffer. When this occurs, the musicians mishear the count-off and risk playing the entire cue out of sync by one beat. A short pre-roll amount (50 ms) may be applied to help to resolve this issue.

29.8 During the Scoring Sessions

Once each DAW session has been prepared specifically for each cue (see Section 28.8), the engineer or DAW operator will need to import the input channels (microphone signals) into each music cue and prepare headphone sends and reverbs, talkback and click-track sends, and any group faders required for quick overall balance adjustments. Having an actual mixing console is a great asset because the setup of peripheral items can remain constant between music cues. Alternatively, these items can be managed within the mixer window of a DAW if it is prepared ahead of time and available in a session template. Here are the steps required, set out as an exercise, assuming Pro Tools is the workstation being used.

EXERCISE 29.2 PREPARING THE DAW TO RECORD EACH CUE

1. Start by building a session that includes all microphones needed. Check that the sample rate and frame rate match the spec provided by the film's production team.
2. Label the tracks as follows; naming might even include an abbreviated version of film title, for example a *Moby Dick* (MD) score would look like this:

   01_MD_1M1_Left main_tk
   02_MD_1M1_Centre main_tk
   03_MD_1M1_Right main_tk

   Numbering the tracks in this way makes it easy to drop files into other sessions directly from the finder, sorted in the correct order rather than alphabetically.

3. Optional: add a live stereo mix track, for reference – although this is less useful when working on an overdub-style score, in which each section or instrument is recorded separately.

4. Make a group called "Record" for all the live inputs and mix tracks, so that it is easy to arm the tracks and to duplicate playlists between takes – ctl/cmd/backslash (\).
5. Be sure to include aux sends for reverb and separate cue sends for click track, prerecords and the live audio mix. Send the click pre-fader so it can be reduced or muted in the control room while remaining at full level in the studio.
6. Set up talkback and listen (conductor) microphones, and if possible, configure these signals so they operate separately from the computer workstation. The conductor's microphone might even be routed to a separate loudspeaker in the control room. That way, proper communication can continue even when the DAW is being closed and opened between music cues and the input signals from the studio are temporarily unavailable.
7. Submasters (built as aux inputs): make sure you have separate "group faders" for your live mics and the prerecorded music stems, and a third one for the dialog/FX audio that came with the picture (called "production audio"). Cue mixes can be sent directly from the submasters as a quick way to add (or remove) certain elements to (or from) the headphone mix.
8. Save this session as a template, so it can be used as a source for all the cues using the *Import Session Data* feature.

The second video output of the computer should be split so that the picture can be displayed in the control room and in the studio or on stage for the conductor, along with the Bars and Beats readout placed below the picture. In film scoring, picture playback is almost more important than the audio! When checking the recording system with the video offline, always be sure to have the picture playback running again before any recording takes place. The director (and anyone else in the control room from the film studio) will be paying attention to the film and how the music affects their impression of the film, rather than specifically listening to the music.

The most important balance during a playback session is between the newly recorded music and the production audio (dialog and FX). The balance should lean toward good musical support of the action that never masks the dialog or any critical sound design. The composer should be watched carefully for their direction in case they want the balance adjusted. The goal is to help convince the director that the music cues contribute effectively throughout the film.

29.9 Notes on the Scoring Session

- Before any recording begins, check for headphone bleed from the click track into the studio.
- Using the playlist function allows all takes to be recorded over the same section of timecode and picture. The recorded take (tk) number will always match the playlist name. You can record a rehearsal in playlist "00" or start off by duplicating the empty playlist so that it advances all record tracks to "Left Main_tk.01", and so on, before recording.
- Just before a new recorded take is made, the playlist should be advanced to the next number. Duplicate Playlist is the recommended shortcut, with any one track of the record group highlighted. This is control, command, backslash (ctl/cmd/backslash) on a Mac. This is better than

using New Playlist, especially when "dropping in" or "punching in" (recording short segments or inserts of a few measures after several complete takes have been recorded).

- Batch Rename is a great shortcut tool for getting all tracks advanced to the correct take number. In the film world, no take number is ever used twice. Every recorded segment is assigned a new take number, with take names going into the hundreds over the course of a full-length film. This is so there are no duplicated filenames, and any missing take can easily be searched for in case it is hidden or was recorded in the wrong session folder. Films engaging multiple engineers, recording studios and music editors can get very messy if proper audio-file housekeeping is overlooked!
- Here is an example of batch track renaming. If a cue is revisited to add some other elements after 33 takes have already been recorded, the playlist number will need to be quickly changed in all the filenames before recording the next take. If enough recording days have gone by since it was last opened, the next available take number might be take 108. To rename, select all the record tracks, then right-click one of the highlighted tracks and scroll down to "Batch Rename". Using the Replace function at the top of the window, enter 33 in Find and 108 in Replace, then hit OK. See Figure 29.5 for an image of the Batch Track Rename window when changing from tk.33 to tk.108.

29.10 Film Score Mixing

When it comes to crafting a surround-sound film score with or without height channels, the main focus or "listening bias" should be toward the front LCR channels. Reverbs, ambiences and other diffuse signals will make up most of the content of the side, rear and height channels, since the audience should not be too distracted by strong music elements localized away from the action on the screen. Certain special-effect components of a mix might work well in the surrounds and, to some extent, in the height channels, such as delayed guitar effects, synth pads or heavily processed piano sounds (chorus, reversed or heavy/dark reverb, etc.).

The center channel should be treated as a significant part of the lateral soundstage, as it helps to anchor the image across a wide listening area. It is good to keep in mind, however, that too much center information may at times conflict with the dialog. This could trigger an overall reduction in the level of the music in the final mix so that the dialog remains clear.

In music mixing for film, it is recommended that the LFE channel be left empty, as it is normally reserved for low-frequency sound effects such as explosions, impacts or thunder. A delivered music LFE channel is most often not used in the final mix, and while at odd times a small amount of music LFE might be desirable, it can be generated from one or more of the main LCR channels for a particular moment in the film.

Music mixes should be checked both with and without the dialog and sound effects running, to make sure everything fits together well in the context of the film mix and that the cues also stand on their own for the soundtrack album release. As opposed to the rerecording mixers, the composer and scoring engineer will be able to introduce more musical level changes as required, in places where changes to a mix may be needed to avoid masking the dialog. There may be moments where only the center channel content of the mix needs to be lowered and there is no need for an overall reduction of level. It is also essential to watch the picture while mixing. Even though most of the balancing will be based on musical parameters and the video might be offline or ignored during the early stages of the mixing process, it is important to watch the picture with the mix running to ensure all the elements in the score are balanced to match the action "on screen".

As mentioned in Section 27.5, running separate reverbs for front, back and center can offer greater control in film mixing. Separate reverbs will improve the quality of the downmix to stereo

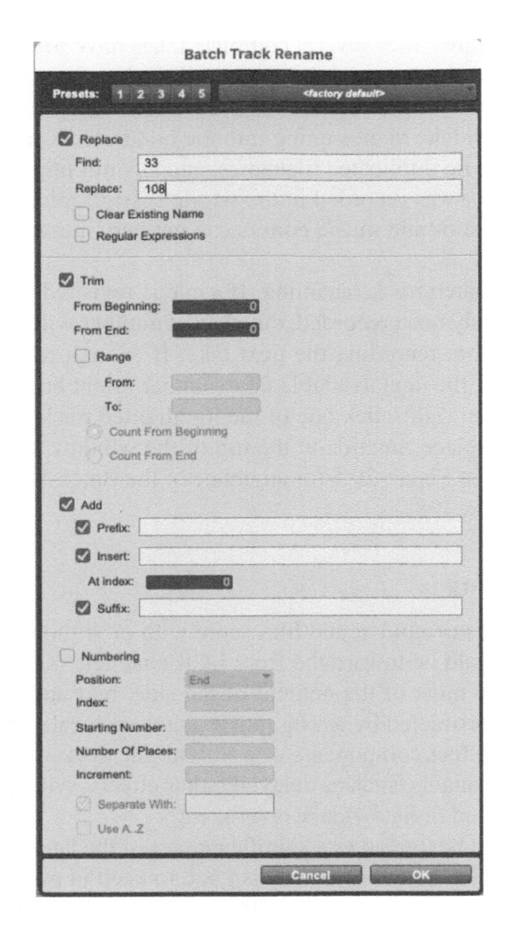

Figure 29.5 Batch Track Rename Window in Pro Tools

Credit: Image courtesy of Avid Technology, Inc.

since the center-channel reverb is a separate "engine" (or iteration) from the left and right reverb channels. While there are a few ways to implement this technique, Figure 29.6 provides one example of how separate mono and stereo reverb engines might be employed in a five-channel film mix. An additional four-channel reverb or two stereo reverbs might be used for height channels in an Atmos immersive film mix.

Endings of music cues that run over into an adjacent scene in a film may need a longer reverb tail in order to match the length of the transition in the picture. This can be quickly achieved by raising the return level through the final decay of the reverb. Short to medium-length reverbs will benefit the most from this technique. Any reverb tails that are deemed too long can easily be shortened in the final mix, so it is good practice to err on the longer side.

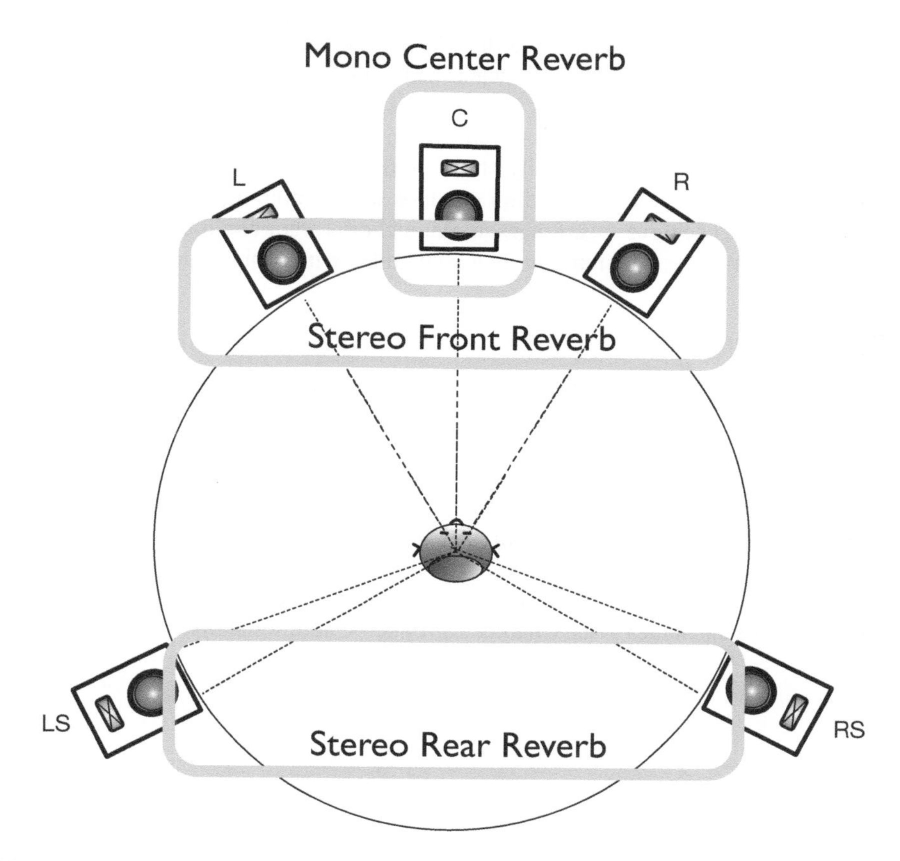

Figure 29.6 Example of Multiple Reverb Engines Employed in a Five-Channel Film Mix

29.11 Printing and Delivering Stems

Stems are separate and isolated mixes of each element in the score. The idea behind delivering stems is that in the final mix the director has the ability to raise, lower or remove any element that was recorded in isolation if it is delivered in its own "stem", including all the associated fader moves, processing and reverb applied. In essence, if all the stems of a cue are set to 0 dB and have the appropriate panning applied, the full score should play back as it was heard by the composer and scoring engineer during the music mix. In most cases, the stems are kept at the original levels during the final mix in order to respect the balances set by the composer and scoring engineer.

Stem "width" may vary, depending on preference and on the importance of each element in the score. For instance, orchestra or string stems may be 5.0 or 7.0 plus 4 height channels, whereas double bass may be printed to a stereo stem. Other less important elements may be LCR only. Proper naming and routing/panning of each stem will guarantee that the mixes will play back properly with all the correct positionings and levels. This is normally carried out and checked by the music editor or the scoring engineer before the files go to the final mix stage.

Small balance changes may be required in certain places throughout the final mix. For example, a cue with sustained strings and some busy percussion may need to be adjusted if it adversely affects a sensitive or soft moment in the dialog. Rather than lowering the entire music mix, it might prove more effective to temporarily reduce the more active percussion stem, while leaving the string stem at normal level if it supports the scene without masking the dialog.

Another advantage of printing stems is that it leaves more options open for the director. They can decide at the last minute if they want to remove an element from the score without affecting anything else in the music mix. Occasionally, the director may ask for an additional music cue during the final mix. With stems, the music editor can look into combining certain instruments across multiple cues in order to achieve this. For example, a chorus stem could be extracted from a cue as an a cappella performance, which might then be blended with some drums from another cue.

29.12 Chapter Summary

- Recording orchestra for film scores is similar to standard classical production, albeit with a few major changes such as perspective and sometimes isolation of sections.
- DAW preparation for each cue is key – this entails synchronizing picture playback, prerecorded files, click track and the Bars and Beats counter (measure numbers) with the printed score.
- A separate DAW "session" is needed for each music start in the film because of the bars and beats and playlist functionalities.
- Immersive film scores such as those mixed for Atmos will require some adaptation during the recording and mixing stages – see Chapter 28 for some additional discussion.

Part VIII

Case Studies

Chapter 30

Recording Orchestra

Case Studies

The following examples come from actual recording sessions or concert recordings involving full orchestra. Recording sessions come with the added stress of the Musician's Union time clock running. Live concert recording generally affords more time for refining the sound and balance during rehearsals; however, the live aspect decrees that once the concert begins, nothing can be adjusted or changed on stage.

30.1 Highly Diffuse Studio Environment

This was a recording titled *Premieres – Cello Concertos*, featuring Yo-Yo Ma (cello) and the Philadelphia Orchestra under David Zinman. Three works were included, by the composers Richard Danielpour, Leon Kirchner and Christopher Rouse. The venue was originally built as a vaudeville-style theater and had been converted into a movie cinema at some point. The space is no longer available as a recording venue, but at the time it was being run as a photography studio for large shoots of products such as automobiles, for example. The room had been stripped of all the theater seats and stage, leaving a large open environment with plaster walls, including some ornate filigree, and a concrete floor that was mostly covered with sheets of plywood.

This case study is interesting because it highlights the differences between human hearing and omnidirectional microphones. The ear and the associated processes of perception and cognition are much more complex than a microphone, which simply reacts to changes in pressure. When standing in a highly diffuse space, we can adjust our auditory system and find clarity and focus in an environment that is otherwise unclear and reverberant. When the same space is auditioned using a pair of omnidirectional microphones, the result will be very different to that of being in the studio with the ensemble. Another disparity when setting microphones is that the recording team normally evaluates the space by standing on stage or sitting in the audience, not several meters, or 10–12 ft., above the podium where the microphones normally reside. During the original setup for this recording, the main system was set at a nominal height just behind the conductor, and the outriggers were positioned similarly, but slightly lower. In most cases, the initial positions are based on experience and are often chosen "by eye".

Once the soundcheck began, it was a real shock to hear the lack of clarity in the mains, and the outriggers were not providing the usual focus on the string section. Short of completely panicking due to lacking the time necessary to refine the sound, a quick decision was made to dramatically change the height of the mains and outriggers. Normally a series of small changes are made, adjustments of a few inches or centimeters until the right balance of focus and reverberation is achieved. In this case, a move of almost 1 meter (3 feet) was implemented in one shot, to bring the direct-to-reverberant balance quickly into check. From that point, the more usual refinements

DOI: 10.4324/9781003319429-38

could be made. I can still remember being amazed at far off we were on the original placement and how surprised we were during the first listening! The room really tricked our ears. Quite a few studio spaces – even a few famous venues –exhibit this same phenomenon to some degree; it is found more often in such spaces than in concert halls.

30.2 Balancing Solo Saxophone with Orchestra

John Adams' Saxophone Concerto was recorded live with Timothy McAllister (alto saxophone) and the St. Louis Symphony, with David Robertson conducting. Powell Hall in St. Louis is a very good hall with a significant amount of reverberation, to the point where the decay is actually better for recording when the audience seating is at least two-thirds full. The initial microphone setup was roughly the same as for *City Noir*, the piece that was coupled with the Saxophone Concerto and which had been recorded the year before in the same hall. The main capture system was double AB, one pair behind the conductor at about 3.65 m or 12 ft., and a second wider AB farther out and up into the hall. The podium was offset to the right to allow the saxophone to be in the center of the stage (see Figure 30.1).

As this recording project was a series of live concerts, the microphone setup was evaluated and adjusted over a few rehearsals leading up to the first concert. The soloist had a very big sound, so much so that the saxophone was quite strong in all the downstage microphones and there was no

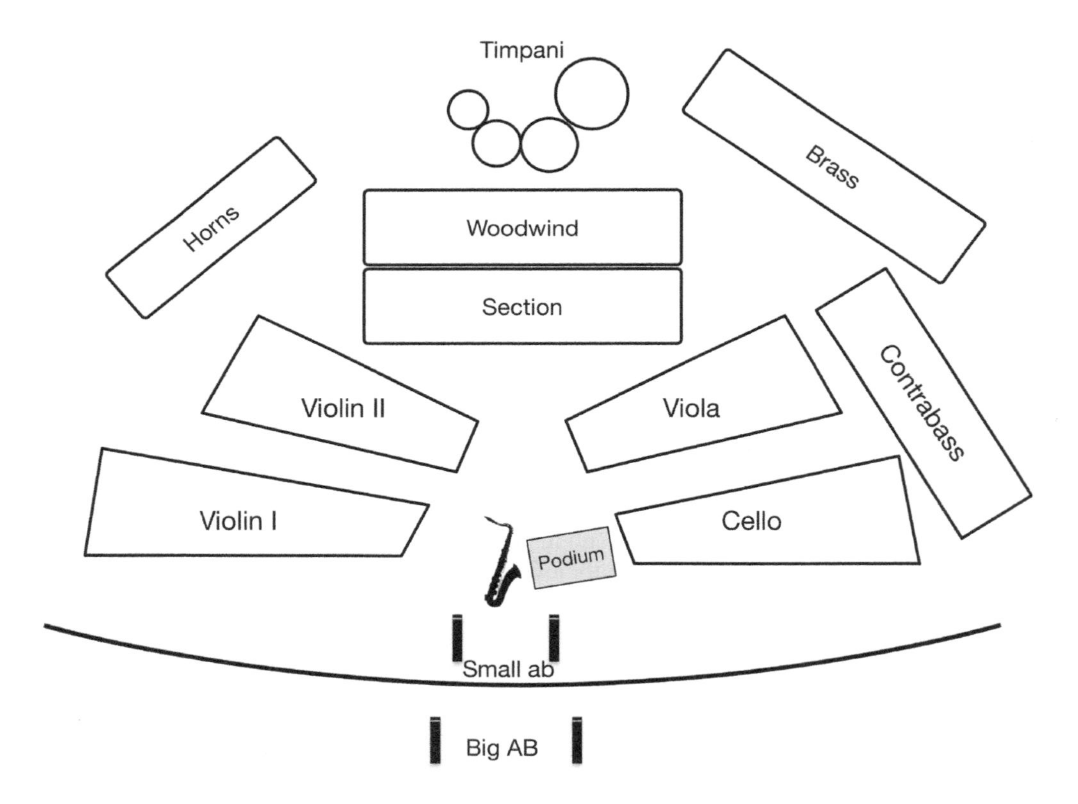

Figure 30.1 Saxophone Concerto Double AB Initial Setup Before Listening

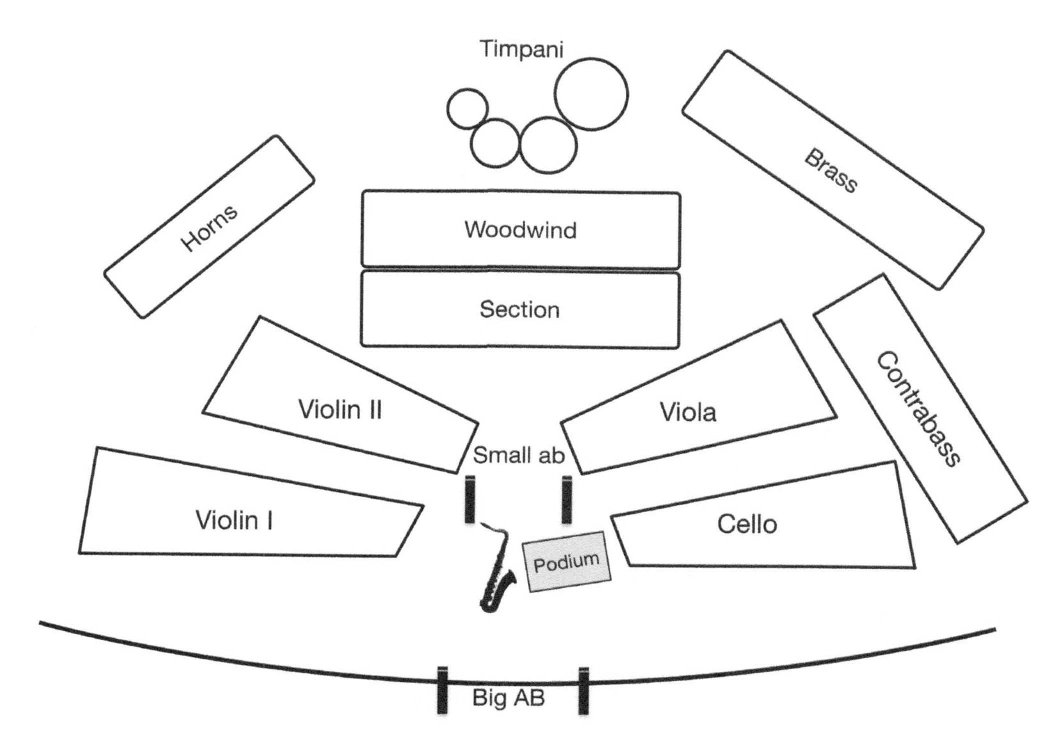

Figure 30.2 Saxophone Concerto Double AB Final Position for Best Balance After Adjusting

way to balance it properly with the strings. It was clear immediately that the main system was not optimized for that piece. As an experiment, the smaller AB was placed farther upstage just behind the soloist, where it might capture more of the strings and less saxophone; we relied on the big AB as the main capture position of the soloist. Of course, there were also microphones for the soloist in place, for clarity and balance. In the end, the small AB ended up quite far upstage during a second major adjustment in order to reduce the amount of saxophone in that pair, while still offering good string coverage in conjunction with the outriggers (see Figure 30.2 – outriggers not shown).

30.3 Faulty Decca Tree Setup

Simple errors occur from time to time in recording sessions, such as left and right main microphones being "swapped" or connected in reverse at some point in the chain. These are normally quite obvious and easy to fix. Add a center main microphone to that, however, and there is more room for error. For instance, left and center may become swapped, creating a "partial" reversing of the image that is less obvious than a full mirroring of the stage sound. Performing a scratch test, as described in Section 14.4, should eliminate any chance of these mix-ups occurring before any listening begins.

Here is a really strange case, however. It was an orchestral session in the legendary Abbey Road Studio One, using a Decca tree of well-maintained Neumann M50 microphones for the main pickup. The M50 houses an omnidirectional capsule mounted in an acrylic sphere that becomes increasingly directional in sensitivity above 1 kHz, transitioning to fully cardioid at around 4 kHz. There is a more detailed discussion of the Decca tree in Chapter 6. A full scratch test was completed during the setup, but when the orchestra arrived and began warming up, the producer noticed a strange "ghosting" in the left channel of instruments sitting on the right of the conductor (cellos in this case). The violins, however, sounded as if they were on the left, although they were slightly dark, and leakage of violin sound into the center and right microphones sounded much brighter. In sum, the image was really distorted, and it wasn't immediately clear why.

With only a few minutes to go before the start of the session, the recording team went out into the studio to have a look. With the Decca tree mounted at more than 3 m (10 feet) it was difficult to tell from the ground what was going on. It did appear however, that the left main microphone had its serial number facing down toward the orchestra rather than the Neumann emblem, which indicates the front face of the microphone. The upside-down placement was quickly confirmed by comparing it with a spare M50 that was on a cart at ground level. A ladder was then rapidly brought in, and the microphone was rotated 180° so that it was properly facing down into the violin section, just one minute before the start of the recording session.

30.4 Mixing Case Study: Violin Concerto

This case study involves an issue with obtaining an artist's approval on a mix. The soloist is not named in this instance, in order to afford full anonymity. A violin concerto was recorded, edited and mixed by the recording team to the satisfaction of the composer, conductor and orchestra. The soloist, however, provided the recording team with several rounds of corrections, both in performance changes (editing) and the solo violin sound. After each iteration of changes the soloist was sent a new edit and mix of the recording with all the requested changes in place.

Finally, with the artist still unhappy with the violin sound, the recording team was at a loss for how they might correct the sound and get the necessary approvals for the release. In the last round, the soloist's comments were sometimes vague or conflicting, to the point that it was impossible to know what they wanted. Descriptions of the desired sound included "more full, more clear, yet more blended and spacious". The producer ultimately said, "let's just raise the violin 1 dB overall, run the mix again and see what they say". This was the magic solution that was needed – after hearing the new mix the soloist was completely satisfied! As it happens, 1 dB of gain on a soloist's level throughout a mix may not seem like a significant change, and it won't really be perceived as being much louder. It will, however, make an instrument sound fuller and clearer and will excite slightly more reverb return when post-fader send mode is used, providing more blend and space. Eureka! Problem solved.

Chapter 31

Recording Chamber Music

Case Studies

This chapter summarizes a few case studies in which issues arose during soundcheck and gives an explanation of how they were solved.

31.1 Situating an Ensemble on the Floor Rather Than the Stage

La Belle Epoque is an album of French repertoire with Yo-Yo Ma (cello) and Kathryn Stott (piano). The recording was made in Mechanics Hall, a beautiful venue in Worcester, Massachusetts, which is excellent for recording chamber music. This hall presents a rather direct, clear and almost dry stage sound while the flat audience area, which is outfitted with removeable, stackable seating, is quite rich and reverberant. In past recordings, the team felt that a disconnect existed between the stage sound and the late-arriving reflections and reverberation of the hall. This phenomenon had to be addressed in postproduction, with some subtle mid-to-late reflection processing to fill the gap.

For this project, a decision was made to try positioning the musicians on the floor area rather than the stage. The idea was to treat the hall as if it were a large recording studio, without the aid of a stage shell to project sound forward into the hall and the microphones. Also, the instruments would be situated in that part of the room where the reflections would arrive earlier and be more present than on the stage area, thereby connecting the direct sound more closely with the reverberant field. The recording team had to commit to this placement, because the piano was to be delivered to the hall during the days leading up to the recording session and there was no option to move the piano up onto the stage area once the delivery was complete.

Once the soundcheck began, the team realized quite quickly that the floor space was much more diffuse than the drier stage area. While the arrival times of the early-to-late reflections were more linear, the level of the reflected sound in relation to the direct sound was much greater than on stage, which then adversely affected the overall clarity. The rich and dense reverb was a positive attribute, but the lack of detail worked against the overall presentation. With no way to move the piano back up onto the stage, the recording team positioned a few makeshift baffles to create side-walls near the musicians and the main microphones, reducing the amount of reflected sound and thereby increasing the definition on the main microphones. This helped to avoid using too much signal from the spot microphones, allowing for a natural perspective in the recording.

The final outcome was quite successful; however, since that project, the team has reverted to positioning the artists on the stage area, prioritizing clarity and detail over the "coupling" of the stage sound with the late reverberation of the hall. In subsequent recordings, the use of a stage extension has helped greatly in connecting the stage sound with the reverberant hall.

DOI: 10.4324/9781003319429-39

31.2 Finding the Optimal Placement for Piano in a Studio

This case study involves Russian repertoire – an album titled *Troika* with Matt Haimovitz (cello) and Christopher O'Riley (piano). The recording took place in the iconic Scoring Stage at Skywalker Sound in Marin County, just north of San Francisco. This is a large studio with variable acoustics that is well suited for both orchestral and chamber music recording. Before the soundcheck, the recording team placed the ensemble so that the piano was near the one-third mark of the length of the room, with the cellist facing the piano. The idea was to avoid the absolute center of the room, and to have a large area of the space free for surround/ambience microphone placement, as the album was planned for release in stereo and five-channel surround as an SACD hybrid disc on the label Pentatone.

After adjusting the microphones for balance, the artists came into the control room to listen. The pianist told us he was hearing a lack of power from the piano, mainly in low frequencies. He also reported that he felt the effect "live" out in the scoring stage and that it was not just a characteristic of the microphone system and the resulting mix. Knowing that the musicians needed to feel comfortable in the studio while performing, the recording team went into the studio with the musicians and moved the piano a few times to try and solve the issue acoustically. The final placement was a few meters or around 10 ft. closer to the rear wall from the original position, even farther away from the center of the room. In this position, the pianist felt the right amount of low-frequency support from the room and good coupling of the instrument with the floor, so that he could perform more comfortably throughout the recording sessions. The actual change in sound as heard over the microphones was less prominent, but there was a decent improvement in the lower range of the piano in terms of overall power, which provided good support for the cello. Here the recording team was able to improve the sound and satisfy the musicians without resorting to the use of equalization on the main system, or other processing after the fact.

Appendix

Quick Start Guides

These tables are designed to provide starting points for recording various types of ensembles in different venues. The tables indicate rough measurements, including height, distance from source and the spacing of an AB pair, presented in a form that is optimized for quick reference. Once the microphones have been positioned according to the following measurements, careful listening will inform the recording team as to how the positions might be adjusted to optimize the overall sound and balance. These tables are designed for use with spaced omnidirectional microphones. In the event that more directional microphones are used, the suggested heights and distances from the source should be increased accordingly. Large Room or Studio assumes a space with no reflectors or reinforcing shell behind the ensemble. Recital Hall refers to a hall of medium reverberation with 200–300 seats, and the Large Hall is based on a fairly reverberant space with 750–1500 seats. The Orchestra Quick Start Guide is also suitable for other large ensembles, such as wind symphony or brass band.

A.1 Orchestra Quick Start Guide

Table A.1 Orchestra Quick Start Guide

ORCHESTRA/LARGE ENSEMBLE QUICK START GUIDE (see Figure A.1 and Chapter 6 for more details, Chapter 16 for wind ensemble and brass band)

PARAMETER	LARGE ROOM OR STUDIO (NO SHELL)	MEDIUM-SIZE RECITAL HALL (200–250 SEATS)	LARGE HALL OR CHURCH (750 + SEATS)
Main Microphone Height	**2.75–2.9 m/9'0"–9'6"**	**3.0 m/10'0"**	**3–3.65 m/10–12'0"**
Microphone Width Apart (a)	**1.5 m/5'0"**	**1.2 m/4'0"**	**1.5 m/5'0"**
Distance Back from Podium (d)	**0 m/0'**, or slightly forward over podium	**30 cm/1'0"**	**0.9 m/3'0"**
Comments	Width (a) depends on ensemble size	Only a small orchestra will fit in a medium hall	Distance (d) may be less for church than large hall

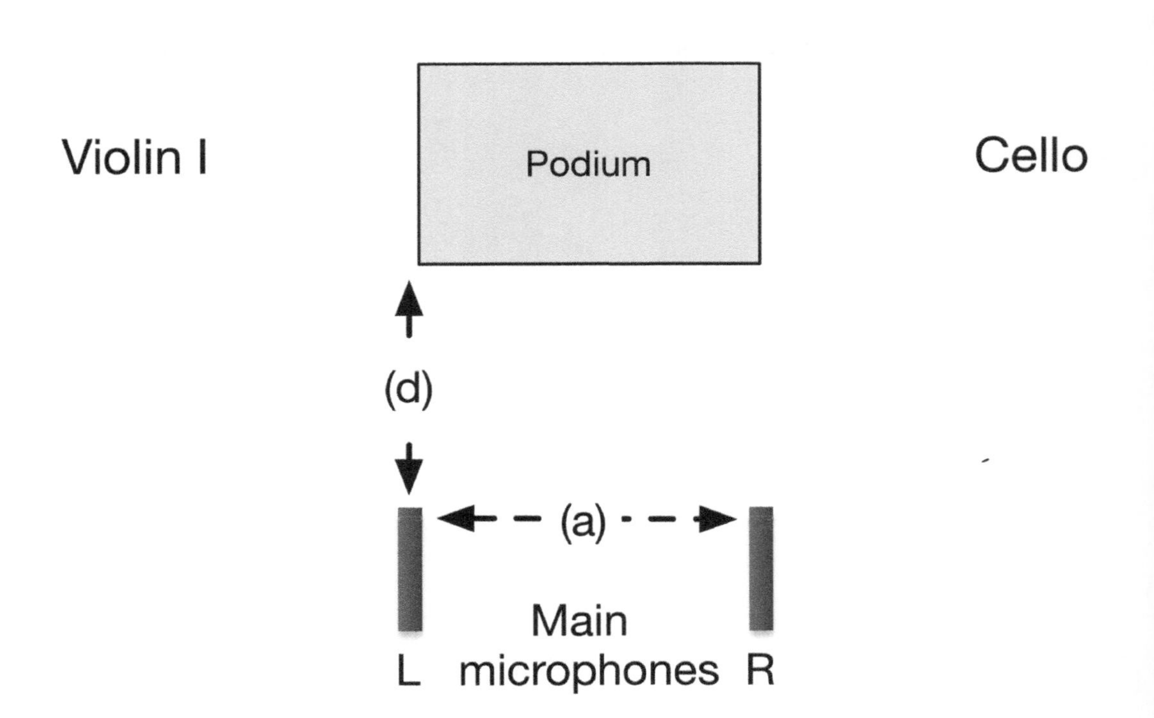

A.2 Chorus Quick Start Guide

Table A.2 Chorus Quick Start Guide

CHORUS QUICK START GUIDE *(see Figure A.2 and Chapter 10 for more details)*			
PARAMETER	*LARGE ROOM OR STUDIO (NO SHELL)*	*MEDIUM-SIZERECITAL HALL (200–250 SEATS)*	*LARGE HALL OR CHURCH (750 + SEATS)*
Main Microphone Height	**2.75 m/9'0"**	**3.0 m/10'0"**	**3.35 m/11'0"**
Microphone Width Apart (a)	**0.9-1.5 m/3–5'0"**	**0.9–1.5 m/3–5'0"**	**0.9–1.5 m/3–5'0"**
Distance Back from Risers (d)	**60 cm/2'0"**	**1.2 m/4'0"**	**1.8 m/6'0"**
Comments	Chorus on risers. Width (a) depends on ensemble size	Chorus on risers. Width (a) depends on ensemble size	Chorus on risers. Width (a) depends on ensemble size

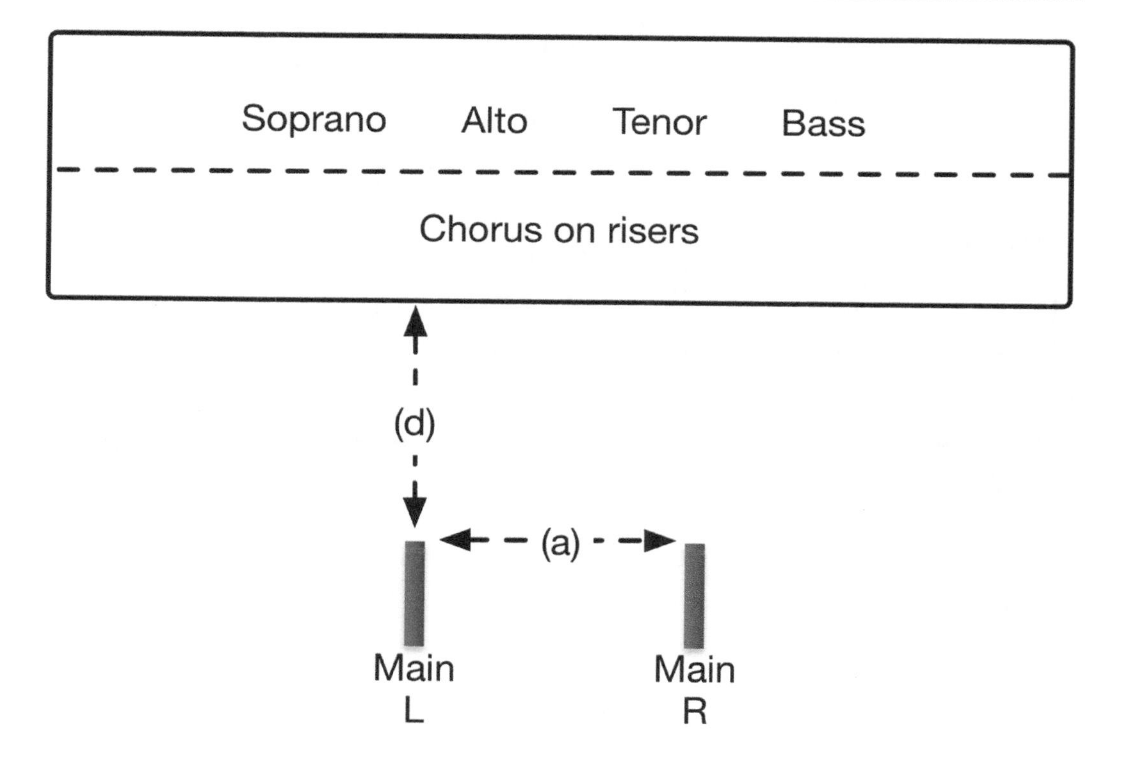

A.3 Chamber Orchestra Quick Start Guide

Table A.3 Chamber Orchestra Quick Start Guide

CHAMBER ORCHESTRA QUICK START GUIDE *(see Figure A.3 and Chapter 17 for more details)*			
PARAMETER	LARGE ROOM OR STUDIO (NO SHELL)	MEDIUM-SIZERECITAL HALL (200–250 SEATS)	LARGE HALL OR CHURCH (750 + SEATS)
Main Microphone Height	L/R: **2.75 m/9'0"** C: **2.6 m/8'6"**	L/R: **2.75–3m/9–10'0"**C: **2.6-2.9 m/8'6"–9'6"**	L/R: **2.75–3m/9-10'0"**C: **2.6–2.9 m/8'6"–9'6"**
L/R Microphone Width Apart (a)	**3.65–4.6 m/12–15'0"**	**3.65–4.6 m/12–15'0"**	**3.65–4.6 m/12–15'0"**
Distance from Podium for Left and Right (d)	**0 m/0'** or slightly forward over strings	**30 cm/1'0"**	**60 cm/2'0"**
Comments	Width (a) depends on ensemble size	Width (a) depends on ensemble size	Width (a) depends on ensemble size

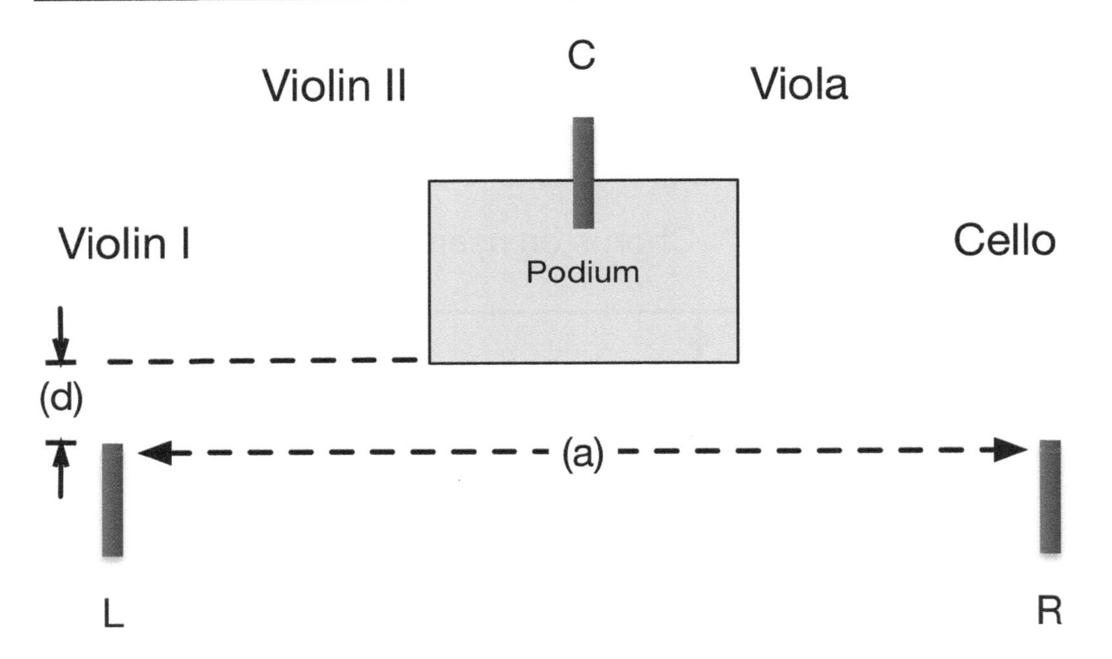

A.4 String Quartet Quick Start Guide

Table A.4 String Quartet Quick Start Guide

STRING QUARTET QUICK START GUIDE *(see Figure A.4 and Chapter 18 for more details)*			
PARAMETER	*LARGE ROOM OR STUDIO (NO SHELL)*	*MEDIUM-SIZERECITAL HALL (200–250 SEATS)*	*LARGE HALL OR CHURCH (750+ SEATS)*
Main Microphone HeightPlacement	**2.6 m/8'6"**	**2.75 m/9'0"**	**3 m/10'0"**
Microphone Width Apart (a)	**70 cm/28"**	**70 cm/28"**	**70 cm/28"**
Distance from Dividing Line of Group (d) – See Diagram	**0–10 cm/0–0'4"**	**10 cm/0'4"**	**20–40cm/8–16"**
Comments	Distance from dividing line affects sense of depth	Distance from dividing line affects sense of depth	Distance from dividing line affects sense of depth

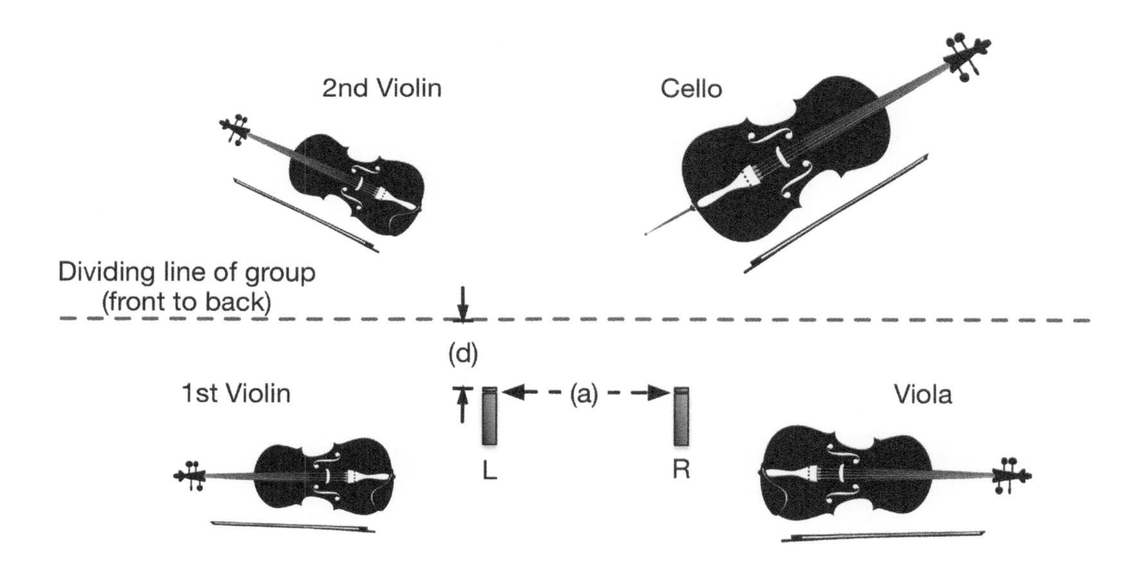

A.5 Woodwind Quintet Quick Start Guide

Table A.5 Woodwind Quintet Quick Start Guide

WOODWIND QUINTET QUICK START GUIDE *(see Figure A.5 and Chapter 19 for more details)*			
PARAMETER	LARGE ROOM OR STUDIO (NO SHELL)	MEDIUM-SIZERECITAL HALL (200–250 SEATS)	LARGE HALL OR CHURCH (750+ SEATS)
Main Microphone HeightPlacement	**2.6 m/8'6"**	**2.75 m/9'0"**	**3 m/10'0"**
Microphone Width Apart (a)	**75 cm/30"**	**75 cm/30"**	**75 cm/30"**
Distance from Front of Group (d) – See Diagram	**50 cm/20"**	**70 cm/28"**	**70 cm/28"**
Comments	Distance from front line affects sense of depth	Distance from front line affects sense of depth	Distance from front line affects sense of depth

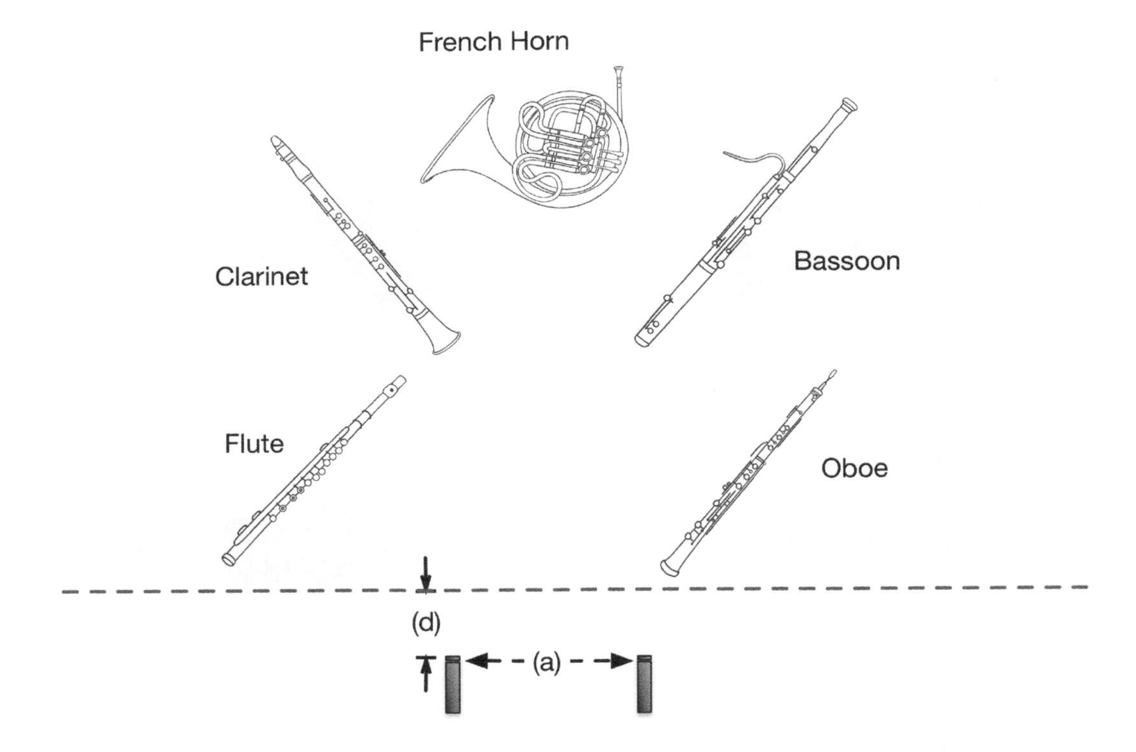

A.6 Brass Quintet Quick Start Guide

Table A.6 Brass Quintet Quick Start Guide

Brass quintet quick start guide (see Figure A.6 and Chapter 19 for more details)			
Parameter	*Large Room or Studio (No Shell)*	*Medium-SizeRecital Hall (200–250 seats)*	*Large Hall or Church (750+ seats)*
Main Microphone HeightPlacement	**2.4 m/8'0"**	**2.6 m/8'6"**	**2.75m/9'0"**
Microphone Width Apart (a)	**80 cm/32"**	**80 cm/32"**	**80 cm/32"**
Distance from Front of Group (d) – See Diagram	**70 cm/28"**	**80 cm/32"**	**90 cm/36"**
Comments	Distance from front line affects sense of depth	Distance from front line affects sense of depth	Distance from front line affects sense of depth

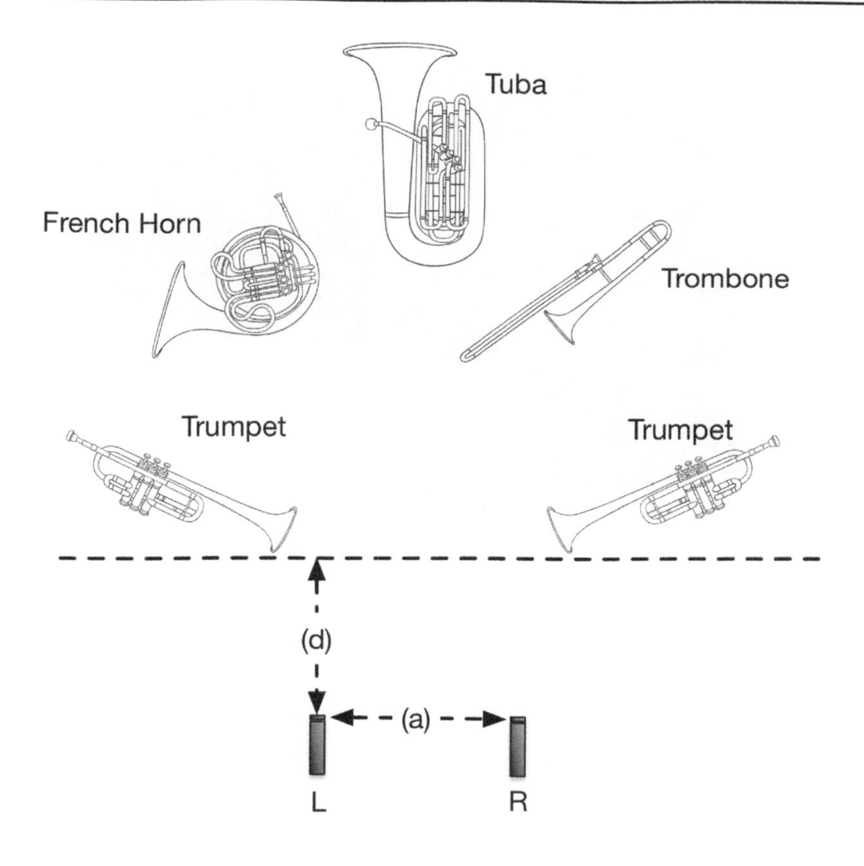

A.7 Piano and Other Instruments Quick Start Guide

Table A.7 Piano with Other Instruments Quick Start Guide

PIANO WITH OTHER INSTRUMENTS QUICK START GUIDE *(see Figure A.7 with alternate positions A and B and Chapter 20 for more details)*			
PARAMETER	LARGE ROOM OR STUDIO (NO SHELL)	MEDIUM-SIZERECITAL HALL (200–250 SEATS)	LARGE HALL OR CHURCH (750+ SEATS)
Main Microphone Height	**2.6 m/8'6"**	**2.6–2.75 m/8'6"–9'0"**	**2.9 m/9'6"**
Microphone Width Apart (a)	**70–80 cm/28–32"**	**70–80 cm/28–32"**	**70–80 cm/28–32"**
A: Distance from Group with Both Instruments Facing Out (d1)	**1–1.4 m/3–4'0"**	**1.4–1.7 m/4–5'0"**	**1.7 m/5'0"**
B: Distance to Piano (d2) and Other Instrument (d3) When Facing Each Other	**d2: 1.5 m/5'0"** **d3: 0.75 m/30"**	**d2: 1.5 m/5'0"** **d3: 0.75 m/30"**	**d2: 1.5 m/5'0"** **d3: 0.75 m/30"**

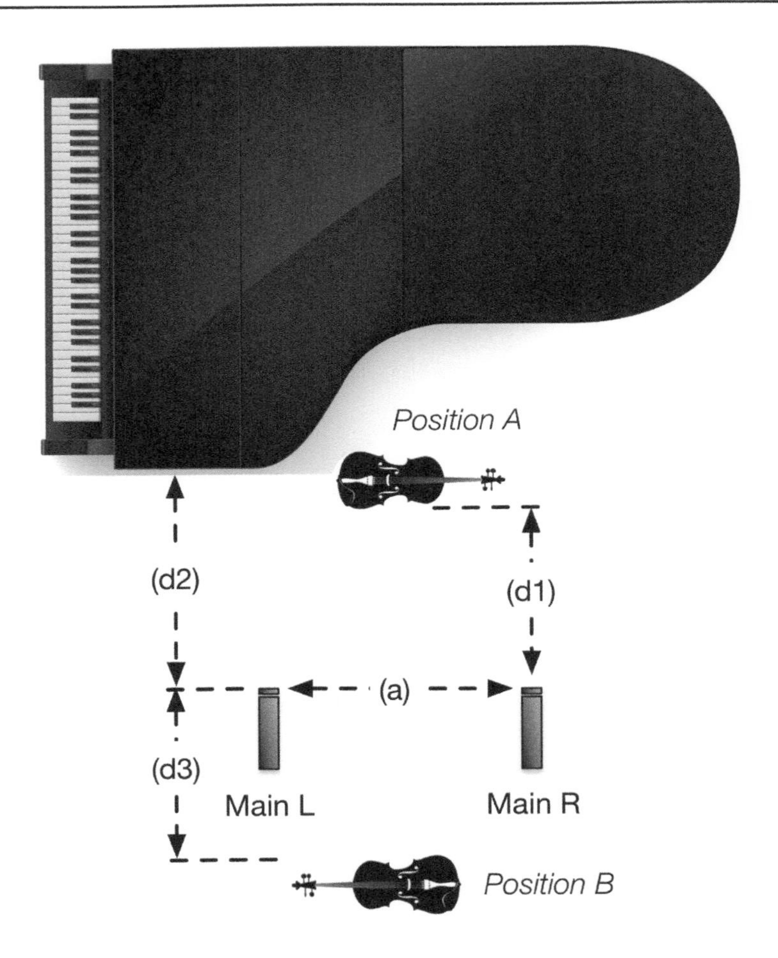

A.8 Solo Piano Quick Start Guide

Table A.8 Solo Piano Quick Start Guide

SOLO PIANO QUICK START GUIDE *(see Figure A.8 and Chapter 21 for more details)*			
PARAMETER	*LARGE ROOM OR STUDIO (NO SHELL)*	*MEDIUM-SIZERECITAL HALL (200–250 SEATS)*	*LARGE HALL OR CHURCH (750+ SEATS)*
Microphone Height	**1.85 m/6'1"**	**1.9 m/6'3"**	**2 m/6'6"**
Microphone Width Apart (a)	**85 cm/34"**	**85 cm/34"**	**85 cm/34"**
Distance from the Outside Hinges (See Figure A.8)	**x1: 2.50 m/98""** **x2: 2.85 m/112"** **y1: 2.60 m/103"** **y2: 2.38 m/95"**	**x1: 2.6 m/102"** **x2: 2.9 m/114"** **y1: 2.7 m/107"** **y2: 2.4 m/96"**	**x1: 2.70 m/106"** **x2: 2.95 m/116"** **y1: 2.80 m/111"** **y2: 2.45 m/97"**
Comments	Piano lid fully open. Width (a) relates to size of instrument	Piano lid fully open. Width (a) relates to size of instrument	Piano lid fully open. Width (a) relates to size of instrument

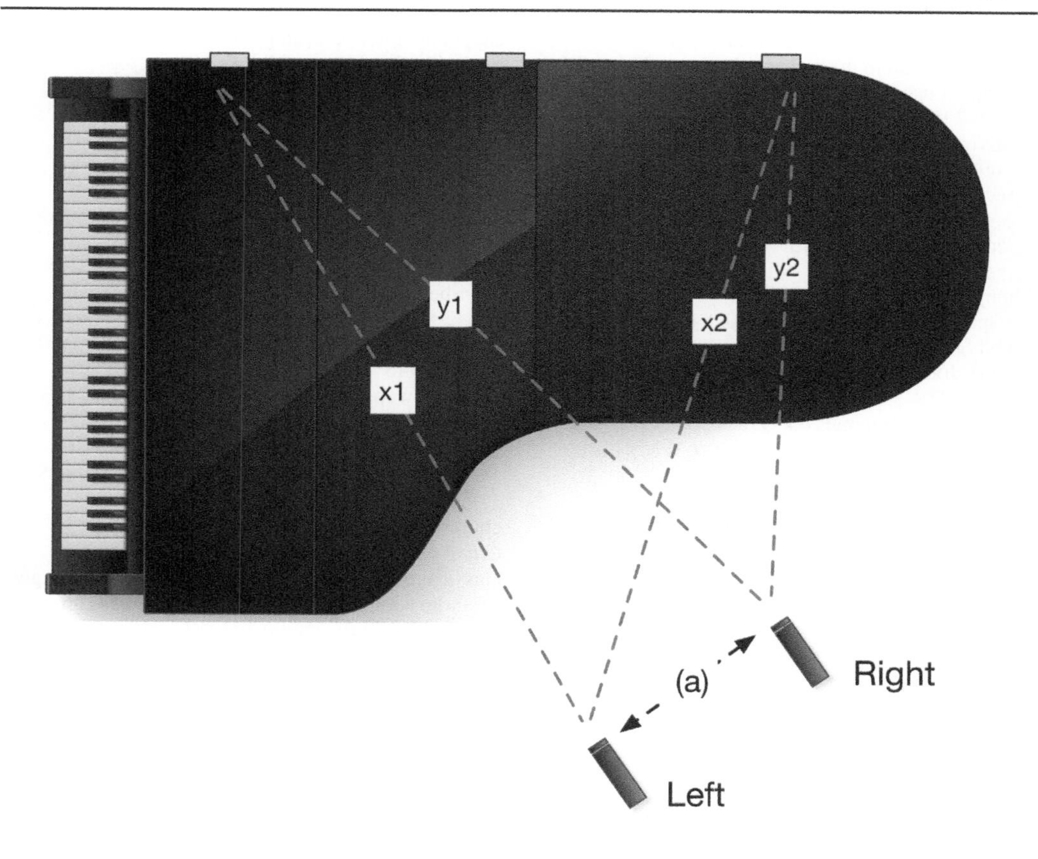

Index

Note: Page numbers in *italics* indicate figures, and page numbers in **bold** indicate tables in the text